Letters on
Natural Magic

Addressed to Sir Walter Scott

DAVID BREWSTER

CAMBRIDGE
UNIVERSITY PRESS

CAMBRIDGE UNIVERSITY PRESS

Cambridge, New York, Melbourne, Madrid, Cape Town, Singapore,
São Paolo, Delhi, Dubai, Tokyo, Mexico City

Published in the United States of America by Cambridge University Press, New York

www.cambridge.org
Information on this title: www.cambridge.org/9781108025553

© in this compilation Cambridge University Press 2010

This edition first published 1832
This digitally printed version 2010

ISBN 978-1-108-02555-3 Paperback

CAMBRIDGE LIBRARY COLLECTION

Books of enduring scholarly value

Spiritualism and Esoteric Knowledge

Magic, superstition, the occult sciences and esoteric knowledge appear regularly in the history of ideas alongside more established academic disciplines such as philosophy, natural history and theology. Particularly fascinating are periods of rapid scientific advances such as the Renaissance or the nineteenth century which also see a burgeoning of interest in the paranormal among the educated elite. This series provides primary texts and secondary sources for social historians and cultural anthropologists working in these areas, and all who wish for a wider understanding of the diverse intellectual and spiritual movements that formed a backdrop to the academic and political achievements of their day. It ranges from works on Babylonian and Jewish magic in the ancient world, through studies of sixteenth-century topics such as Cornelius Agrippa and the rapid spread of Rosicrucianism, to nineteenth-century publications by Sir Walter Scott and Sir Arthur Conan Doyle. Subjects include astrology, mesmerism, spiritualism, theosophy, clairvoyance, and ghost-seeing, as described both by their adherents and by sceptics.

Letters on Natural Magic

Intended as a supplement to Sir Walter Scott's 1830 *Letters on Demonology and Witchcraft*, this 1832 publication seeks to explain and expose the science behind the alleged 'magic' of spiritualists and conjurors. David Brewster (1781–1868), a Scottish natural philosopher and historian of science, was highly regarded in his lifetime but has since faded into obscurity. Penned at the request of Scott, Brewster's friend and neighbour, this book follows an epistolary structure, consisting of thirteen letters each addressing and exposing different aspects of the alleged supernatural activity, in keeping with the format of Scott's publication. Brewster's subject matter includes optics, magic lanterns, automata, alchemy, fire-breathing, spontaneous combustion, spectral illusions and various other phenomena. In each case he carefully outlines how this 'magic' is created with optical illusion, narcotic drugs, gas inhalation, and chemical tricks. The book offers an intriguing insight into nineteenth-century attitudes towards the supernatural.

Cambridge University Press has long been a pioneer in the reissuing of out-of-print titles from its own backlist, producing digital reprints of books that are still sought after by scholars and students but could not be reprinted economically using traditional technology. The Cambridge Library Collection extends this activity to a wider range of books which are still of importance to researchers and professionals, either for the source material they contain, or as landmarks in the history of their academic discipline.

Drawing from the world-renowned collections in the Cambridge University Library, and guided by the advice of experts in each subject area, Cambridge University Press is using state-of-the-art scanning machines in its own Printing House to capture the content of each book selected for inclusion. The files are processed to give a consistently clear, crisp image, and the books finished to the high quality standard for which the Press is recognised around the world. The latest print-on-demand technology ensures that the books will remain available indefinitely, and that orders for single or multiple copies can quickly be supplied.

The Cambridge Library Collection will bring back to life books of enduring scholarly value (including out-of-copyright works originally issued by other publishers) across a wide range of disciplines in the humanities and social sciences and in science and technology.

LETTERS

ON

NATURAL MAGIC,

ADDRESSED TO

Sir WALTER SCOTT, Bart.

BY

Sir DAVID BREWSTER, K. H.

LL.D. F.R.S. V.P.R.S.E. &c. &c.

———

LONDON:

JOHN MURRAY, ALBEMARLE STREET.

MDCCCXXXII.

CONTENTS.

LETTER III.

LETTER IV.

LETTER V.

LETTER VI.

LETTER VII.

LETTER VIII.

LETTER IX.

LETTER X.

LETTER XI.

LETTER XII.

LETTER XIII.

CONTENTS.

LETTERS

ON

NATURAL MAGIC,

ADDRESSED TO

Sir WALTER SCOTT, Bart.

LETTER I.

Extent and interest of the subject—Science employed by ancient governments to deceive and enslave their subjects—Influence of the supernatural upon ignorant minds—Means employed by the ancient magicians to establish their authority—Derived from a knowledge of the phenomena of Nature—From the influence of narcotic drugs upon the victims of their delusion—From every branch of science—Acoustics — Hydrostatics — Mechanics—Optics — M. Salverte's work on the occult sciences—Object of the following Letters.

My Dear Sir Walter,

As it was at your suggestion that I undertook to draw up a popular account of those prodigies of the material world which have received the appella-

tion of *Natural Magic*, I have availed myself of
the privilege of introducing it under the shelter of
your name.   Although I cannot hope to produce
a volume at all approaching in interest to that
which you have contributed to the Family Library,
yet the popular character of some of the topics
which belong to this branch of Demonology may
atone for the defects of the following Letters ; and
I shall deem it no slight honour if they shall be
considered as forming an appropriate supplement
to your valuable work.

The subject of Natural Magic is one of great
extent as well as of deep interest.   In its widest
range, it embraces the history of the governments
and the superstitions of ancient times,—of the
means by which they maintained their influence
over the human mind,—of the assistance which
they derived from the arts and the sciences, and
from a knowledge of the powers and phenomena of
nature.   When the tyrants of antiquity were unable or unwilling to found their sovereignty on
the affections and interests of their people, they
sought to entrench themselves in the strongholds
of supernatural influence, and to rule with the de-
legated authority of heaven.   The prince, the priest,
and the sage, were leagued in a dark conspiracy to
deceive and enslave their species; and man, who refused his submission to a being like himself, became
the obedient slave of a spiritual despotism, and willingly bound himself in chains when they seemed
to have been forged by the gods.

This system of imposture was greatly favoured
by the ignorance of these early ages.   The human
mind is at all times fond of the marvellous, and the

3

credulity of the individual may be often measured by his own attachment to the truth. When knowledge was the property of only one caste, it was by no means difficult to employ it in the subjugation of the great mass of society. An acquaintance with the motions of the heavenly bodies, and the variations in the state of the atmosphere, enabled its possessor to predict astronomical and meteorological phenomena with a frequency and an accuracy which could not fail to invest him with a divine character. The power of bringing down fire from the heavens, even at times when the electric influence was itself in a state of repose, could be regarded only as a gift from heaven. The power of rendering the human body insensible to fire was an irresistible instrument of imposture ; and in the combinations of chemistry, and the influence of drugs and soporific embrocations on the human frame, the ancient magicians found their most available resources.

The secret use which was thus made of scientific discoveries and of remarkable inventions, has no doubt prevented many of them from reaching the present times ; but though we are very ill informed respecting the progress of the ancients in various departments of the physical sciences, yet we have sufficient evidence that almost every branch of knowledge had contributed its wonders to the magician's budget, and we may even obtain some insight into the scientific acquirements of former ages, by a diligent study of their fables and their miracles.

The science of *Acoustics* furnished the ancient sorcerers with some of their best deceptions. The imitation of thunder in their subterranean temples could not fail to indicate the presence of a supernatural agent. The golden virgins whose ravishing voices resounded through the temple of Delphos;—the stone from the river Pactolus, whose trumpet notes scared the robber from the treasure which it guarded;—the speaking head which uttered its oracular responses at Lesbos;—and the vocal statue of Memnon, which began at the break of day to accost the rising sun,—were all deceptions derived from science, and from a diligent observation of the phenomena of nature.

The principles of *Hydrostatics* were equally available in the work of deception. The marvellous fountain which Pliny describes in the Island of Andros as discharging wine for seven days, and water during the rest of the year;—the spring of oil which broke out in Rome to welcome the return of Augustus from the Sicilian war,—the three empty urns which filled themselves with wine at the annual feast of Bacchus in the city of Elis,—the glass tomb of Belus which was full of oil, and which, when once emptied by Xerxes, could not again be filled,—the weeping statues, and the perpetual lamps of the ancients,—were all the obvious effects of the equilibrium and pressure of fluids.

Although we have no direct evidence that the philosophers of antiquity were skilled in *Mechanics*, yet there are indications of their knowledge, by no means equivocal, in the erection of the Egyptian obelisks, and in the transportation of huge masses of stone, and their subsequent elevation to great heights in their temples. The powers which they employed, and the mechanism by which they operated, have

been studiously concealed, but their existence may
be inferred from results otherwise inexplicable, and
the inference derives additional confirmation from
the mechanical arrangements which seem to have
formed a part of their religious impostures. When
in some of the infamous mysteries of ancient Rome,
the unfortunate victims were carried off by the gods,
there is reason to believe that they were hurried
away by the power of machinery ; and when Apollonius, conducted by the Indian sages to the temple
of their god, felt the earth rising and falling beneath his feet like the agitated sea, he was no
doubt placed upon a moving floor capable of imitating the heavings of the waves. The rapid descent
of those who consulted the oracle in the cave of
Trophonius,—the moving tripods which Apollonius saw in the Indian temples,—the walking statues at Antium, and in the Temple of Hierapolis,—
and the wooden pigeon of Archytas, are specimens
of the mechanical resources of the ancient magic.

But of all the sciences *Optics* is the most fertile
in marvellous expedients. The power of bringing
the remotest objects within the very grasp of the
observer, and of swelling into gigantic magnitude the almost invisible bodies of the material
world, never fails to inspire with astonishment even
those who understand the means by which these
prodigies are accomplished. The ancients, indeed,
were not acquainted with those combinations of
lenses and mirrors which constitute the telescope
and the microscope, but they must have been familiar with the property of lenses and mirrors to
form erect and inverted images of objects. There
is reason to think that they employed them to ef-

fect the apparition of their gods ; and in some of the descriptions of the optical displays which hallowed their ancient temples, we recognize all the transformations of the modern phantasmagoria.

It would be an interesting pursuit to embody the information which history supplies respecting the fables and incantations of the ancient superstitions, and to show how far they can be explained by the scientific knowledge which then prevailed. This task has, to a certain extent, been performed by M. Eusebe Salverte, in a work on the occult sciences, which has recently appeared ; but notwithstanding the ingenuity and learning which it displays, the individual facts are too scanty to support the speculations of the author, and the descriptions are too meagre to satisfy the curiosity of the reader.*

In the following letters I propose to take a wider range, and to enter into more minute and popular details. The principal phenomena of nature, and the leading combinations of art, which bear the impress of a supernatural character, will pass under our review, and our attention will be particularly called to those singular illusions of sense, by which the most perfect organs either cease to perform their functions, or perform them faithlessly ; and where the efforts and the creations of the mind predominate over the direct perceptions of external nature.

---

* We must caution the young reader against some of the views given in M. Salverte's work. In his anxiety to account for every thing miraculous by natural causes, he has ascribed to the same origin some of those events in sacred history which Christians cannot but regard as the result of divine agency.

In executing this plan, the task of selection is rendered extremely difficult, by the superabundance of materials, as well as from the variety of judgments for which these materials must be prepared. Modern science may be regarded as one vast miracle, whether we view it in relation to the Almighty Being, by whom its objects and its laws were formed, or to the feeble intellect of man, by which its depths have been sounded, and its mysteries explored :—And if the philosopher who is familiarized with its wonders, and who has studied them as necessary results of general laws, never ceases to admire and adore their author, how great should be their effect upon less gifted minds, who must ever view them in the light of inexplicable prodigies.— Man has in all ages sought for a sign from heaven, and yet he has been habitually blind to the millions of wonders with which he is surrounded. If the following pages should contribute to abate this deplorable indifference to all that is grand and sublime in the universe, and if they should inspire the reader with a portion of that enthusiasm of love and gratitude which can alone prepare the mind for its final triumph, the labours of the author will not have been wholly fruitless.

LETTER II.

*The Eye the most important of our organs—Popular descrip-
tion of it—The eye is the most fertile source of mental illusions—Disappearance of objects when their images fall upon
the base of the optic nerve—Disappearance of objects when
seen obliquely—Deceptions arising from viewing objects in a
faint light—Luminous figures created by pressure on the eye
either from external causes or from the fulness of the blood-
vessels—Ocular spectra or accidental colours—Remarkable
effects produced by intense light—Influence of the imagination in viewing these Spectra—Remarkable illusion produced
by this affection of the eye—Duration of impressions of light
on the eye—Thaumatrope—Improvements upon it suggested—Disappearance of halves of objects or of one of two persons—Insensibility of the eye to particular colours—Remarkable optical illusion described.*

OF all the organs by which we acquire a knowledge of external nature the eye is the most remarkable and the most important. By our other
senses the information we obtain is comparatively
limited. The touch and the taste extend no fur-
ther than the surface of our own bodies. The sense
of smell is exercised within a very narrow sphere, and
that of recognizing sounds is limited to the distance
at which we hear the bursting of a meteor and the
crash of a thunderbolt. But the eye enjoys a bound-
less range of observation. It takes cognizance not
only of other worlds belonging to the solar system,

but of other systems of worlds infinitely removed into the immensity of space ; and when aided by the telescope, the invention of human wisdom, it is able to discover the forms, the phenomena, and the movements of bodies whose distance is as inexpressible in language as it is inconceivable in thought.

While the human eye has been admired by ordinary observers for the beauty of its form, the power of its movements, and the variety of its expression, it has excited the wonder of philosophers by the exquisite mechanism of its interior, and its singular adaptation to the variety of purposes which it has to serve. The eye-ball is nearly globular, and is about an inch in diameter. It is formed externally by a tough opaque membrane called the *sclerotic* coat, which forms the white of the eye, with the exception of a small circular portion in front called the *cornea*. This portion is perfectly transparent, and so tough in its nature as to afford a powerful resistance to external injury. Immediately within the cornea, and in contact with it, is the *aqueous* humour, a clear fluid, which occupies only a small part of the front of the eye. Within this humour is the *iris*, a circular membrane with a hole in its centre called the pupil. The colour of the eye resides in this membrane, which has the curious property of contracting and expanding so as to diminish or enlarge the pupil,—an effect which human ingenuity has not been able even to imitate. Behind the iris is suspended the *crystalline* lens in a fine transparent capsule or bag of the same form with itself. It is then succeeded by the *vitreous humour*, which resembles the transparent white of an egg, and fills up the rest of the eye. Behind the vitreous humour, there is spread out on the inside of the eye-ball a fine delicate membrane, called the *retina,* which is an expansion of the *optic nerve,* entering the back of the eye, and communicating with the brain.

A perspective view and horizontal section of the left eye, shown in the annexed figure, will convey a popular idea of its structure. It is, as it were, a small camera obscura, by means of which the pictures of external objects are painted on the retina, and in a way of which we are ignorant, it conveys the impression of them to the brain.

Fig. 1.

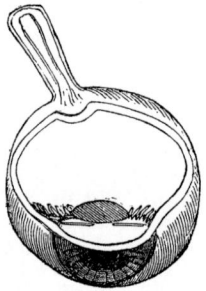

This wonderful organ may be considered as the sentinel which guards the pass between the worlds of matter and of spirit, and through which all their communications are interchanged. The optic nerve is the channel by which the mind peruses the handwriting of Nature on the retina, and through which it transfers to that material tablet its decisions and its creations. The eye is consequently the principal seat of the supernatural. When the indications of the marvellous are addressed to us through the ear, the mind may be startled without being deceived, and reason may succeed in suggesting some probable source of the illusion by which we have been alarmed : But when the eye in solitude sees before it the forms of life, fresh in their colours and vivid in their outline ; when distant or departed friends are suddenly presented to its view ; when visible bodies disappear and reappear without any intelligible cause ; and when it beholds objects, whether real or imaginary, for whose presence no cause can be assigned, the conviction of supernatural agency becomes under ordinary circumstances unavoidable.

Hence it is not only an amusing but an useful occupation to acquire a knowledge of those causes which are capable of producing so strange a belief, whether it arises from the delusions which the mind practises upon itself, or from the dexterity and science of others. I shall therefore proceed to explain those illusions which have their origin in the eye, whether they are general, or only occasionally exhibited in particular persons, and under particular circumstances.

There are few persons aware that when they look with one eye there is some particular object before them to which they are absolutely blind. If we look with the right eye this point is always about 15° to the right of the object which we are viewing, or to the right of the axis of the eye or the point of most distinct vision. If we look with the left eye the point is as far to the left. In order to be convinced of this curious fact, which was discovered by M. Mariotte, place two coloured wafers upon a sheet of white paper at the distance of three inches, and look at the left hand wafer with the right eye at the distance of about 11 or 12 inches, taking care to keep the eye straight above the wafer, and the line which joins the eyes parallel to the line which joins the wafers. When this is done, and the left eye closed, the right hand wafer will no longer be visible. The same effect will be produced if we close the right eye and look with the left eye at the right hand wafer. When we examine the retina to discover to what part of it this insensibility to light belongs, we find that the image of the invisible wafer has fallen on the base of the optic nerve, or the place where this nerve enters the eye and expands itself to form the retina. This point is shown in the preceding figure by a convexity at the place where the nerve enters the eye.

But though light of ordinary intensity makes no impression upon this part of the eye, a very strong light does, and even when we use candles or highly luminous bodies in place of wafers the body does not wholly disappear, but leaves behind a faint cloudy light, without, however, giving any thing like an image of the object from which the light proceeds.

When the objects are *white* wafers upon a *black* ground, the white wafer absolutely disappears, and the space which it covers appears to be completely black; and as the light which illuminates a landscape is not much different from that of a white wafer, we should expect, whether we use one or both eyes,* to see a black or a dark spot upon every land-

---

* When both eyes are open, the object whose image falls upon the insensible spot of the one eye is seen by the other, so that though it is not invisible, yet it will only be half as luminous, and therefore two dark spots ought to be seen.

scape, within 15° of the point which most particularly attracts our notice. The Divine Artificer,
however, has not left his work thus imperfect.
Though the base of the optic nerve is insensible to
light that falls directly upon it, yet it has been made
susceptible of receiving luminous impressions from
the parts which surround it, and the consequence
of this is, that when the wafer disappears, the spot
which it occupied, in place of being black, has always
the same colour as the ground upon which the wafer
is laid, being white when the wafer is placed upon
a white ground, and red when it is placed upon a
red ground.   This curious effect may be rudely illustrated by comparing the retina to a sheet of blot-
ting-paper, and the base of the optic nerve to a circular portion of it covered with a piece of sponge.
If a shower falls upon the paper, the protected part
will not be wetted by the rain which falls upon the
sponge that covers it, but in a few seconds it will
be as effectually wetted by the moisture which it
absorbs from the wet paper with which it is surrounded.   In like manner the insensible spot on
the retina is stimulated by a borrowed light, and
the apparent defect is so completely removed that
its existence can be determined only by the experiment already described.

Of the same character, but far more general in
its effects, and important in its consequences, is
another illusion of the eye which presented itself to me several years ago.   When the eye is
steadily occupied in viewing any particular object,
or when it takes a fixed direction while the mind
is occupied with any engrossing topic of speculation
or of grief, it suddenly loses sight of, or becomes

blind to, objects seen indirectly, or upon which it is not fully directed. This takes place whether we use one or both eyes, and the object which disappears will reappear without any change in the position of the eye, while other objects will vanish and revive in succession without any apparent cause. If a sportsman, for example, is watching with intense interest the motions of one of his dogs, his companion, though seen with perfect clearness by indirect vision, will vanish, and the light of the heath or of the sky will close in upon the spot which he occupied.

In order to witness this illusion, put a little bit of white paper on a green cloth, and within three or four inches of it, place a narrow strip of white paper. At the distance of twelve or eighteen inches, fix one eye steadily upon the little bit of white paper, and in a short time a part or even the whole of the strip of paper will vanish as if it had been removed from the green cloth. It will again reappear, and again vanish, the effect depending greatly on the steadiness with which the eye is kept fixed. This illusion takes place when both the eyes are open, though it is easier to observe it when one of them is closed. The same thing happens when the object is luminous. When a candle is thus seen by indirect vision, it never wholly disappears, but it spreads itself out into a cloudy mass, the centre of which is blue, encircled with a bright ring of yellow light.

This inability of the eye to preserve a sustained vision of objects seen obliquely, is curiously compensated by the greater sensibility of those parts of the eye that have this defect. The eye has the power of seeing objects with perfect distinctness, only when it is directed straight upon them; that is, all objects seen indirectly are seen indistinctly; but it is a curious circumstance, that when we wish to obtain a sight of a very faint star, such as one of the satellites of Saturn, we can see it most distinctly *by looking away from it*, and when the eye is turned full upon it, it immediately disappears.

Effects still more remarkable are produced in the eye when it views objects that are difficult to be seen from the small degree of light with which they happen to be illuminated. The imperfect view which we obtain of such objects forces us to fix the eye more steadily upon them; but the more exertion we make to ascertain what they are, the greater difficulties do we encounter to accomplish our object. The eye is actually thrown into a state of the most painful agitation, the object will swell and contract, and partly disappear, and it will again become visible when the eye has recovered from the delirium into which it has been thrown. This phenomenon may be most distinctly seen when the objects in a room are illuminated with the feeble gleam of a fire almost extinguished; but it may be observed in day-light by the sportsman when he endeavours to mark upon the monotonous heath the particular spot where moor-game has alighted. Availing himself of the slightest difference of tint in the adjacent heath, he keeps his eye steadily fixed on it as he advances, but whenever the contrast of illumination is feeble, he will invariably lose sight of his mark, and if the retina is capable of taking it up, it is only to lose it a second time.

This illusion is likely to be most efficacious in

the dark, when there is just sufficient light to render white objects faintly visible, and to persons who
are either timid or credulous must prove a frequent
source of alarm. Its influence too is greatly aided by
another condition of the eye, into which it is thrown
during partial darkness. The pupil expands nearly
to the whole width of the iris in order to collect the
feeble light which prevails; but it is demonstrable
that in this state the eye cannot accommodate itself
to see near objects distinctly, so that the form of
persons and things actually become more shadowy
and confused when they come within the very distance at which we count upon obtaining the best
view of them. These affections of the eye are, we
are persuaded, very frequent causes of a particular
class of apparitions which are seen at night by the
young and the ignorant. The spectres which are
conjured up are always *white*, because no other colour can be seen, and they are either formed out of
inanimate objects which reflect more light than
others around them, or of animals or human beings
whose colour or change of place renders them more
visible in the dark. When the eye dimly descries
an inanimate object whose different parts reflect
different degrees of light, its brighter parts may enable the spectator to keep up a continued view of
it; but the disappearance and reappearance of its
fainter parts, and the change of shape which ensues,
will necessarily give it the semblance of a living
form, and if it occupies a position which is unap-
proachable, and where animate objects cannot find
their way, the mind will soon transfer to it a supernatural existence. In like manner a human figure
shadowed forth in a feeble twilight may undergo

similar changes, and after being distinctly seen while it is in a situation favourable for receiving and reflecting light, it may suddenly disappear in a position fully before, and within the reach of, the observer's eye; and if this evanescence takes place in a path or road where there was no side-way by which the figure could escape, it is not easy for an ordinary mind to efface the impression which it cannot fail to receive.   Under such circumstances, we never think of distrusting an organ which we have never found to deceive us; and the truth of the maxim that " seeing is believing," is too universally admitted, and too deeply rooted in our nature to admit on any occasion of a single exception.

In these observations we have supposed that the spectator bears along with him no fears or prejudices, and is a faithful interpreter of the phenomena presented to his senses ; but if he is himself a believer in apparitions, and unwilling to receive an ocular demonstration of their reality, it is not difficult to conceive the picture which will be drawn when external objects are distorted and caricatured by the imperfect indications of his senses, and coloured with all the vivid hues of the imagination.

Another class of ocular deceptions have their origin in a property of the eye which has been very imperfectly examined.   The fine nervous fabric which constitutes the retina, and which extends to the brain, has the singular property of being *phosphorescent by pressure*.   When we press the eye-ball outwards by applying the point of the finger between it and the nose, a circle of light will be seen, which Sir Isaac Newton describes as " a circle of colours like those in the feather of a peacock's tail."   H e

B

adds, that " if the eye and the finger remain quiet,
these colours vanish in a second of time, but if the
finger be moved with a quavering motion they appear again." In the numerous observations which
I have made on these luminous circles, I have
never been able to observe any colour but white,
with the exception of a general red tinge which
is seen when the eyelids are closed, and which is
produced by the light which passes through them.
The luminous circles too always continue while the
pressure is applied, and they may be produced as
readily after the eye has been long in darkness as
when it has been recently exposed to light. When
the pressure is very gently applied, so as to compress the fine pulpy substance of the retina, light
is immediately created when the eye is in total
darkness ; and when in this state light is allowed
to fall upon it, the part compressed is more sensible to light than any other part, and consequently
appears more luminous. If we increase the pressure, the eye-ball, being filled with incompressible
fluids, will protrude all round the point of pressure,
and consequently the retina at the protruded part
will be *compressed* by the outward pressure of the
contained fluid, while the retina on each side, namely, under the point of pressure and beyond the protruded part, will be drawn towards the protruded
part or *dilated*. Hence the part under the finger
which was originally compressed is now *dilated*,
the adjacent parts *compressed*, and the more remote
parts immediately without this *dilated* also. Now
we have observed, that when the eye is, under these
circumstances, exposed to light, there is a bright
luminous circle shading off externally and inter-

nally into total darkness. We are led therefore to the
important conclusions, that when the retina is compressed in total darkness it gives out light ; that
when it is compressed when exposed to light, its
sensibility to light is increased ; and that when it is
*dilated under exposure to light, it becomes absolute-
ly blind, or insensible to all luminous impressions.*

When the body is in a state of perfect health,
this phosphorescence of the eye shows itself on
many occasions. When the eye or the head receives a sudden blow, a bright flash of light shoots
from the eyeball. In the act of sneezing, gleams
of light are emitted from each eye, both during
the inhalation of the air, and during its subsequent protrusion, and in blowing air violently
through the nostrils, two patches of light appear above the axis of the eye and in front of it,
while other two luminous spots unite into one, and
appear as it were about the point of the nose when
the eyes are directed to it. When we turn the
eyeball by the action of its own muscles, the retina is affected at the place where the muscles are
inserted, and there may be seen opposite each
eye and towards the nose, two semicircles of light,
and other two extremely faint towards the temples.
At particular times, when the retina is more phosphorescent than at others, these semicircles are expanded into complete circles of light.

In a state of indisposition, the phosphorescence
of the retina appears in new and more alarming
forms. When the stomach is under a temporary derangement accompanied with headach, the pressure
of the blood-vessels upon the retina shows itself,
in total darkness, by a faint blue light floating be-

fore the eye, varying in its shape, and passing
away at one side. This blue light increases in intensity, becomes *green* and then *yellow*, and sometimes rises to *red*, all these colours being frequently seen at once, or the mass of light shades off into
darkness. When we consider the variety of distinct forms which in a state of perfect health the
imagination can conjure up when looking into a burning fire, or upon an irregularly shaded surface, * it is
easy to conceive how the masses of coloured light
which float before the eye may be moulded by the
same power into those fantastic and natural shapes
which so often haunt the couch of the invalid, even
when the mind retains its energy, and is conscious
of the illusion under which it labours. In other
cases, temporary blindness is produced by pressure
upon the optic nerve, or upon the retina, and under the excitation of fever or delirium, when the

---

* A very curious example of the influence of the imagination in creating distinct forms out of an irregularly shaded surface, is mentioned in the life of Peter Heaman, a Swede,
who was executed for piracy and murder at Leith in 1822.
We give it in his own words:
"One remarkable thing was, one day as we mended a sail,
it being a very thin one, after laying it upon deck in folds,
I took the tar brush and tarred it over in the places which I
thought needed to be strengthened. But when we hoisted
it up, I was astonished to see that the tar I had put upon it
represented a gallows and a man under it without a head.
The head was lying beside him. He was complete, body,
thighs, legs, arms, and in every shape like a man. Now, I
oftentimes made remarks upon it, and repeated them to the
others. I always said to them all, you may depend upon it
that something will happen. I afterwards took down the sail
on a calm day, and sewed a piece of canvass over the figure
to cover it, for I could not bear to have it always before
my eyes."

physical cause which produces spectral forms is at
its height, there is superadded a powerful influence
of the mind, which imparts a new character to the
phantasms of the senses.

In order to complete the history of the illusions
which originate in the eye, it will be necessary to
give some account of the phenomena called *ocular
spectra*, or *accidental colours*.   If we cut a figure
out of red paper, and placing it on a sheet of white
paper, view it steadily for some seconds with one
or both eyes fixed on a particular part of it, we
shall observe the red colour to become less brilliant.
If we then turn the eye from the red figure upon
the white paper, we shall see a distinct *green* figure, which is the *spectrum*, or accidental colour
of the *red* figure.   With differently coloured figures we shall observe differently coloured spectra,
as in the following table :

Colour of the Original figures.	Colour of the Spectral figures.
Red,	Bluish-green.
Orange,	Blue.
Yellow,	Indigo.
Green,	Reddish-violet.
Blue,	Orange-red.
Indigo,	Orange-yellow.
Violet,	Yellow.
White,	Black.
Black,	White.

The two last of these experiments, viz. white and
black figures, may be satisfactorily made by using
a white medallion on a dark ground, and a black
profile figure.   The spectrum of the former will be
found to be black, and that of the latter white.

These ocular spectra often show themselves without any effort on our part, and even without our knowledge. In a highly painted room illuminated by the sun, those parts of the furniture on which the sun does not directly fall, have always the opposite or accidental colour. If the sun shines through a chink in a *red* window-curtain, its light will appear *green*, varying, as in the above table, with the colour of the curtain ; and if we look at the image of a candle reflected from the water in a *blue* finger glass, it will appear *yellow*. Whenever, in short, the eye is affected with one prevailing colour, it sees at the same time the spectral or accidental colour, just as when a musical string is vibrating, the ear hears at the same time its fundamental and its harmonic sounds.

If the prevailing light is *white* and *very strong*, the spectra which it produces are no longer black, but of various colours in succession. If we look at the sun, for example, when near the horizon, or when reflected from glass or water, so as to moderate its brilliancy, and keep the eye upon it steadily for a few seconds, we shall see even for hours afterwards, and whether the eyes is open or shut, a spectre of the sun varying in its colours. At first, with the eye open, it is *brownish-red* with a *sky-blue* border, and when the eye is shut, it is *green* with a *red* border. The *red* becomes more brilliant, and the *blue* more vivid, till the impression is gradually worn off ; but even when they become very faint, they may be revived by a gentle pressure on the eyeball.

Some eyes are more susceptible than others of these spectral impressions, and Mr Boyle mentions

an individual who continued for years to see the
spectre of the sun when he looked upon bright objects. This fact appeared to Locke so interesting
and inexplicable, that he consulted Sir Isaac Newton respecting its cause, and drew from him the
following interesting account of a similar effect
upon himself: " The observation you mention in
Mr Boyle's book of colours, I once made upon
myself with the hazard of my eyes. The manner
was this : I looked a very little while upon the
sun in the looking-glass with my right eye, and
then turned my eyes into a dark corner of my chamber, and winked, to observe the impression made.
and the circles of colours which encompassed it,
and how they decayed by degrees, aud at last vanished. This I repeated a second and a third time.
At the third time, when the phantasm of light and
colours about it were almost vanished, intending
my fancy upon them to see their last appearance,
I found, to my amazement, that they began to
return, and by little and little to become as lively
and vivid as when I had newly looked upon the
sun. But when I ceased to intend my fancy upon
them they vanished again. After this, I found,
that, as often as I went into the dark, and intended my mind upon them, as when a man looks ear-
nestly to see any thing which is difficult to be seen,
I could make the phantasm return without looking
any more upon the sun ; and the oftener I made it
return, the more easily I could make it return
again. And at length, by repeating this without
looking any more upon the sun, I made such an
impression on my eye, that, if I looked upon the
clouds, or a book, or any bright object, I saw upon

it a round bright spot of light like the sun, and, which is still stranger, though I looked upon the sun with my right eye only, and not with my left, yet my fancy began to make an impression upon my left eye as well as upon my right.   For if I shut my right eye, and looked upon a book or the clouds with my left eye, I could see the spectrum of the sun almost as plain as with my right eye, if I did but intend my fancy a little while upon it ; for at first, if I shut my right eye, and looked with my left, the spectrum of the sun did not appear till I intended my fancy upon it ; but by repeating, this appeared every time more easily.   And now in a few hours time I had brought my eyes to such a pass, that I could look upon no bright object with either eye but I saw the sun before me, so that I durst neither write nor read ; but to recover the use of my eyes, shut myself up in my chamber made dark, for three days together, and used all means in my power to direct my imagination from the sun.   For if I thought upon him, I presently saw his picture, though I was in the dark.   But by keeping in the dark, and employing my mind about other things, I began in three or four days to have more use of my eyes again ; and by forbearing to look upon bright objects, recovered them pretty well ; though not so well but that, for some months after, the spectrums of the sun began to return as often as I began to meditate upon the phenomena, even though I lay in bed at midnight with my curtains drawn.   But now I have been very well for many years, though I am apt to think, if I durst venture my eyes, I could still make the phantasm return by the power of my fancy.   This story I tell you, to let you understand, that in the observation related by Mr Boyle, the man's fancy probably concurred with the impression made by the sun's light to produce that phantasm of the sun which he constantly saw in bright objects." *

I am not aware of any effects that had the character of supernatural having been actually produced by the causes above described ; but it is obvious, that, if a living figure had been projected against the strong light which imprinted these durable spectra of the sun, which might really happen when the solar rays are reflected from water, and diffused by its ruffled surface, this figure would have necessarily accompanied all the luminous spectres which the fancy created.   Even in ordinary lights strange appearances may be produced by even transient impressions, and if I am not greatly mistaken, the case which I am about to mention is not only one which may occur, but which actually happened.   A figure dressed in *black* and mounted upon a *white* horse was riding along exposed to the bright rays of the sun, which through a small opening in the clouds was throwing its light only upon that part of the landscape.   The *black* figure was projected against a white cloud, and the white horse shone with particular brilliancy by its contrast with the dark soil against which it was seen.   A person interested in the arrival of such a stranger had been for some time following his movements with intense anxiety, but upon his disappearance behind a wood, was surprised to observe the spectre of the mounted stranger in the form of a *white* rider upon a *black* steed, and this spectre was seen for

* See the *Edinburgh Encyclopædia*, Art. ACCIDENTAL COLOURS.

some time in the sky, or upon any pale ground to
which the eye was directed.   Such an occurrence,
especially if accompanied with a suitable combination of events, might even in modern times have
formed a chapter in the history of the marvellous.

It is a curious circumstance, that when the image
of an object is impressed upon the retina only for
a few moments, the picture which is left is exactly
of the same colour with the object. If we look, for
example, at a window at some distance from the eye,
and then transfer the eye quickly to the wall, we
shall see it distinctly but momentarily with *light*
panes and *dark* bars ; but in a space of time incal-
culably short, this picture is succeeded by the spectral impression of the window, which will consist of
black panes and *white* bars. The similar spectrum, or
that of the same colour as the object, is finely seen
in the experiment of forming luminous circles by
whirling round a burning stick, in which case the
circles are always red.

In virtue of this property of the eye an object
may be seen in many places at once ; and we may
even exhibit at the same instant the two opposite
sides of the same object, or two pictures painted on
the opposite sides of a piece of card.   It was found
by a French philosopher M. D'Arcet, that the impression of light continued on the retina about the
eighth part of a second after the luminous body was
withdrawn, and upon this principle Dr Paris has
constructed the pretty little instrument, called the
Thaumatrope, or the *Wonder-turner*.   It consists
of a number of circular pieces of card about two or
three inches broad, which may be twirled round with
great velocity by the application of the fore-finger
and thumb of each hand to pieces of silk string at-

tached to opposite points of their circumference.
On each side of the circular piece of card is painted part of a picture, or a part of a figure, in such a
manner that the two parts would form a group or
a whole figure if we could see both sides at once.
Harlequin for example is painted on one side, and
Columbine on the other, so that by twirling round
the card the two are seen at the same time in their
usual mode of combination. The body of a Turk is
drawn on one side, and his head on the reverse, and
by the rotation of the card the head is replaced upon
his shoulders. The principle of this illusion may be
extended to many other contrivances. Part of a sen-
tence may be written on one side of a card and the
rest on the reverse. Particular letters may be given
on one side, and others upon the other, or even
halves or parts of each letter may be put upon each
side, or all these contrivances may be combined,
so that the sentiment which they express can be
understood only when all the scattered parts are
united by the revolution of the card.

As the revolving card is virtually transparent,
so that bodies beyond it can be seen through it,
the power of the illusion might be greatly extended by introducing into the picture other figures,
either animate or inaminate. The setting sun for
example, might be introduced into a landscape: Part
of the flame of a fire might be seen to issue from
the crater of a volcano, and cattle grazing in a field
might make part of the revolutionary landscape.
For such purposes, however, the form of the instrument would require to be completely changed, and
the rotation should be effected round a standing
axis by wheels and pinions, and a screen placed in

front of the revolving plane with open compartments or apertures, through which the principal figures would appear. Had the principle of this instrument been known to the ancients, it would doubtless have formed a powerful engine of delusion in their temples, and might have been more effective than the optical means which they seem to have employed for producing the apparitions of their gods.

In certain diseased conditions of the eye effects of a very remarkable kind are produced. The faculty of seeing objects double is too common to be noticed as remarkable; and though it may take place with only one eye, yet, as it generally arises from a transient inability to direct the axes of both eyes to the same point, it excites little notice. That state of the eye, however, in which we lose sight of half of every object at which we look, is more alarming and more likely to be ascribed to the disappearance of part of the object than to a defect of sight. Dr Wollaston, who experienced this defect twice, informs us that, after taking violent exercise, he " suddenly found that he could see but half of a man whom he met, and that, on attempting to read the name of JOHNSON over a door, he saw only SON, the commencement of the name being wholly obliterated from his view." In this instance, the part of the object which disappeared was towards his left, but on a second occurrence of the same affection, the part which disappeared was towards his right. There are many occasions on which this defect of the eye might alarm the person who witnessed it for the first time. At certain distances from the eye one of two persons would necessarily disappear; and by a slight change of position either in the observer or the person observed, the person that vanished would reappear, while the other would disappear in his turn. The circumstances under which these evanescences would take place could not be supposed to occur to an ordinary observer, even if he should be aware that the cause had its origin in himself. When a phenomenon so strange is seen by a person in perfect health, as it generally is, and who has never had occasion to distrust the testimony of his senses, he can scarcely refer it to any other cause than a supernatural one.

Among the affections of the eye which not only deceive the person who is subject to them, but those also who witness their operation, may be enumerated the insensibility of the eye to particular colours. This defect is not accompanied with any imperfection of vision, or connected with any disease either of a local or a general nature, and it has hitherto been observed in persons who possess a strong and a sharp sight. Mr Huddart has described the case of one Harris, a shoemaker at Maryport in Cumberland, who was subject to this defect in a very remarkable degree. He seems to have been insensible to every colour, and to have been capable of recognizing only the two opposite tints of *black* and *white*. " His first suspicion of this defect arose when he was about four years old. Having by accident found in the street a child's stocking, he carried it to a neighbouring house to inquire for the owner: He observed the people call it a *red stocking*, though he did not understand why they gave it that denomination, as he himself thought it completely described by being called a stocking. The circumstance, however, remained in

his memory, and, with other subsequent observations, led him to the knowledge of his defect. He
observed also that, when young, other children
could discern cherries on a tree by some pretended
difference of colour, though he could only distinguish them from the leaves by their difference of
size and shape. He observed also that, by means
of this difference of colour, they could see the cherries at a greater distance than he could, though he
could see other objects at as great a distance as they,
that is, where the sight was not assisted by the colour." Harris had two brothers whose perception
of colours was nearly as defective as his own. One
of these, whom Mr Huddart examined, constantly
mistook *light green* for *yellow*, and *orange* for *grass
green.*

Mr Scott has described in the Philosophical
Transactions his own defect in perceiving colours. He states that he does not know any *green*
in the world; that a *pink* colour and a *pale blue*
are perfectly alike ; that he has often thought a
full red and a *full green* a good match ; that he is
sometimes baffled in distinguishing a *full purple*
from a *deep blue,* but that he knows light, dark, and
middle *yellows,* and all degrees of *blue* except *sky-
blue.* " I married my daughter to a genteel, worthy
man, a few years ago; the day before the marriage he
came to my house dressed in a new suit of fine cloth
clothes. I was much displeased that he should come
as I supposed, in *black* ; and said that he should go
back to change his colour. But my daughter said,
No—No ; the colour is very genteel, that it was
my eyes that deceived me. He was a gentleman
of the law, in a fine rich claret-coloured dress, which

is as much a black to my eyes as any black that ever was dyed." Mr Scott's father, his maternal uncle, one of his sisters, and her two sons, had all the same imperfection. Dr Nichol has recorded a case where a naval officer purchased a *blue* uniform coat and waistcoat with *red* breeches to match the blue, and Mr Harvey describes the case of a tailor at Plymouth, who on one occasion repaired an article of dress with *crimson* in place of *black* silk, and on another patched the elbow of a *blue* coat with a piece of *crimson* cloth. It deserves to be remarked that our celebrated countrymen the late Mr Dugald Stewart, Mr Dalton, and Mr Troughton, have a similar difficulty in distinguishing colours. Mr Stewart discovered this defect when one of his family was admiring the beauty of the Siberian crab-apple, which he could not distinguish from the leaves but by its form and size. Mr Dalton cannot distinguish *blue* from pink, and the solar spectrum consists only of two colours, *yellow* and *blue*. Mr Troughton regards *red*, *ruddy pinks* and brilliant *oranges*, as *yellows*, and *greens* as *blues*, so that he is capable only of appreciating *blue* and *yellow* colours.

In all those cases which have been carefully studied, at least in three of them in which I have had the advantage of making personal observations, namely, those of Mr Troughton, Mr Dalton, and Mr Liston, the eye is capable of seeing the whole of the prismatic spectrum, the red space appearing to be yellow. If the red space consisted of homogeneous or simple red rays, we should be led to infer that the eyes in question were not insensible to red light, but were merely incapable of discriminating between the impressions of red and yellow light. I have lately shown, however, that the prismatic spectrum consists of three equal and coincident spectra of *red, yellow,* and *blue* light, and consequently, that much yellow and a small portion of blue light exist in the red space;—and hence it follows, that those eyes which see only two colours, viz. *yellow* and *blue,* in the spectrum, are really insensible to the red light of the spectrum, and see only the yellow with the small portion of blue with which the red is mixed. The faintness of the yellow light which is thus seen in the red space, confirms the opinion that the retina has not appreciated the influence of the simple red rays.

If one of the two travellers who, in the fable of the chameleon, are made to quarrel about the colour of that singular animal, had happened to possess this defect of sight, they would have encountered at every step of their journey new grounds of dissension without the chance of finding an umpire who could pronounce a satisfactory decision. Under certain circumstances, indeed, the arbiter might set aside the opinions of both the disputants, and render it necessary to appeal to some higher authority

—— to beg he'd tell 'em if he knew
Whether the thing was *red* or *blue.*

In the course of writing the preceding observations, an ocular illusion occurred to myself of so extraordinary nature, that I am convinced it never was seen before, and I think it far from probable that it will ever be seen again. Upon directing my eyes to the candles that were standing before me, I was surprised to observe, apparently among my hair, and nearly straight above my head, and far without the range of vision, a distinct image of one of the candles inclined about 45° to the horizon; as shown at A in Fig. 2.   The image was as distinct and

Fig. 2.

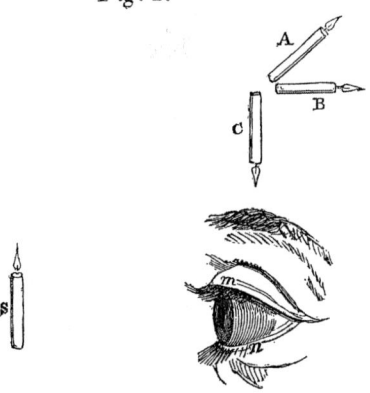

perfect as if it had been formed by reflexion from a piece of mirror glass, though of course much less brilliant, and the position of the image proved that it must be formed by reflexion from a perfectly flat and highly polished surface.   But where such a surface could be placed, and how, even if it were fixed, it could reflect the image of the candle up through my head, were difficulties not a little perplexing. Thinking that it might be something lodged in the eyebrow, I covered it up from the light, but the image still retained its place.   I then examined the eyelashes with as little success, and was driven to the extreme supposition that a crystallization was taking place in some part of the aqueous humour of the eye, and that the image was formed by the

c

reflexion of the light of the candle from one of the crystalline faces.   In this state of uncertainty, and, I may add, of anxiety, for this last supposition was by no means an agreeable one, I set myself down to examine the phenomenon experimentally. I found that the image varied its place by the motion of the head and of the eyeball, which proved that it was either attached to the eyeball or occupied a place where it was affected by that motion.   Upon inclining the candle at different angles the image suffered corresponding variations of position. In order to determine the exact place of the reflecting substance, I now took an opaque circular body and held it between the eye and the candle till it eclipsed the mysterious image.   By bringing the body nearer and nearer the eyeball till its shadow became sufficiently distinct to be seen, it was easy to determine the locality of the reflector, because the shadow of the opaque body must fall upon it whenever the image of the candle was eclipsed. In this way I ascertained that the reflecting body was in the upper eyelash, and I found, that, in consequence of being disturbed, it had twice changed its inclination, so as to represent a vertical candle in the horizontal position B, and afterwards in the inverted position C.   Still, however, I sought for it in vain, and even with the aid of a magnifier I could not discover it. At last, however, Mrs B. who possesses the perfect vision of short-sighted persons, discovered, after repeated examinations, between two eyelashes, a minute speck, which, upon being removed with great difficulty, turned out to be a chip of red wax not above the hundredth part of an inch in diameter, and having its surface so perfectly

4

flat and so highly polished that I could see in it
the same image of the candle, by placing it extremely near the eye.   This chip of wax had no
doubt received its flatness and its polish from the
surface of a seal, and had started into my eye when
breaking the seal of a letter.

That this reflecting substance was the cause of
the image of the candle cannot admit of a doubt ;
but the wonder still remains how the images which
it formed occupied so mysterious a place as to be
seen without the range of vision, and apparently
through the head.   In order to explain this, let
m n, Fig. 2, be a lateral view of the eye.   The chip
of wax was placed at m at the root of the eyelashes,
and being nearly in contact with the outer surface
of the cornea, the light of the candle which it reflected passed very obliquely through the pupil and fell
upon the retina somewhere to the left of n, very near
where the retina terminates ; but a ray thus falling
obliquely on the retina is seen, in virtue of the law
of visible direction already explained, in a line n C
perpendicular to the retina at the point near n, where
the ray fell.   Hence the candle was necessarily seen
through the head as it were of the observer, and
without the range of ordinary vision.   The compa-
rative brightness of the reflected image still surprises
me ; but even this, if the image really was brighter,
may be explained by the fact, that it was formed on
a part of the retina, upon which light had never before fallen, and which may therefore be supposed
to be more sensible, than the parts of the membrane
in constant use, to luminous impressions.

Independent of its interest as an example of the
marvellous in vision, the preceding fact may be

considered as a proof that the retina retains its power to its very termination near the ciliary processes, and that the law of visible direction holds true even without the range of ordinary vision. It is therefore possible that a reflecting surface favourably placed on the outside of the eye, or that a reflecting surface in the inside of the eye, may cause a luminous image to fall nearly on the extreme margin of the retina, the consequence of which would be that it would be seen in the back of the head half way between a vertical and a horizontal line.

LETTER III.

*Subject of spectral illusions—Recent and interesting case of
Mrs A.—Her first illusion affecting the ear—Spectral apparition of her husband—Spectral apparition of a cat—
Apparition of a near and living relation in grave-clothes
seen in a looking-glass—Other illusions affecting the ear
—Spectre of a deceased friend sitting in an easy chair
—Spectre of a coach and four filled with skeletons—Accuracy and value of the preceding cases—State of health under
which they arose—Spectral apparitions are pictures on the
retina—The ideas of memory and imagination are also
pictures on the retina—General views of the subject—Approximate explanation of spectral apparitions.*

THE preceding account of the different sources of
illusion to which the eye is subject, is not only useful as indicating the probable cause of any individual deception, but it has a special importance in
preparing the mind for understanding those more
vivid and permanent spectral illusions to which some
individuals have been either occasionally or habitually subject.

In these lesser phenomena we find the retina so
powerfully influenced by external impressions as to
retain the view of visible objects long after they are
withdrawn ; we observe it to be so excited by local
pressures of which we sometimes know neither the
nature nor the origin, as to see in total darkness mov-

ing and shapeless masses of coloured light ; and we find, as in the case of Sir Isaac Newton and others, that the imagination has the power of reviving the impressions of highly luminous objects, months and even years after they were first made. From such phenomena, the mind feels it to be no violent transition to pass to those spectral illusions which, in particular states of health, have haunted the most intelligent individuals, not only in the broad light of day, but in the very heart of the social circle.

This curious subject has been so ably and fully treated in your Letters on Demonology, that it would be presumptuous in me to resume any part of it on which you have even touched; but as it forms a necessary branch of a Treatise on Natural Magic, and as one of the most remarkable cases on record has come within my own knowledge, I shall make no apology for giving a full account of the different spectral appearances which it embraces, and of adding the results of a series of observations and experiments on which I have been long occupied, with the view of throwing some light on this remarkable class of phenomena.

A few years ago I had occasion to spend some days under the same roof with the lady to whose case I have above referred. At that time she had seen no spectral illusions, and was acquainted with the subject only from the interesting volume of Dr Hibbert. In conversing with her about the cause of these apparitions, I mentioned, that, if she should ever see such a thing, she might distinguish a genuine ghost existing externally, and seen as an external object, from one created by the mind, by merely pressing one eye or straining them both so as to see objects double ; for in this case the external object or supposed apparition would invariably be doubled, while the impression on the retina created by the mind would remain single. This observation recurred to her mind when she unfortunately became subject to the same illusions ; but she was too well acquainted with their nature to require any such evidence of their mental origin ; and the state of agitation which generally accompanies them seems to have prevented her from making the experiment as a matter of curiosity.

1. The first illusion to which Mrs A. was subject was one which affected only the ear. On the 26th of December 1830, about half-past four in the afternoon, she was standing near the fire in the hall and on the point of going up stairs to dress, when she heard, as she supposed, her husband's voice calling her by name, " —— ——Come here! come to me!" She imagined that he was calling at the door to have it opened, but upon going there and opening the door she was surprised to find no person there. Upon returning to the fire, she again heard the same voice calling out very distinctly and loudly, " —— Come, come here!" She then opened two other doors of the same room, and upon seeing no person she returned to the fire-place. After a few moments she heard the same voice still calling, " —— ——Come to me, come! come away!" in a loud, plaintive, and somewhat impatient tone. She answered as loudly, "Where are you? I don't know where you are ;" still imagining that he was somewhere in search of her: but receiving no answer she shortly went up stairs. On Mr A.'s return to the house, about half an hour afterwards, she inquired

why he called to her so often, and where he was; and she was of course greatly surprised to learn that he had not been near the house at the time. A similar illusion, which excited no particular notice at the time, occurred to Mrs A. when residing at Florence about ten years before, and when she was in perfect health. When she was undressing after a ball, she heard a voice call her repeatedly by name, and she was at that time unable to account for it.

2. The next illusion which occurred to Mrs A. was of a more alarming character. On the 30th of December, about four o'clock in the afternoon, Mrs A. came down stairs into the drawing-room, which she had quitted only a few minutes before, and on entering the room she saw her husband, as she supposed, standing with his back to the fire. As he had gone out to take a walk about half an hour before, she was surprised to see him there, and asked him why he had returned so soon. The figure looked fixedly at her with a serious and thoughtful expression of countenance but did not speak. Supposing that his mind was absorbed in thought, she sat down in an arm chair near the fire, and within two feet at most of the figure, which she still saw standing before her. As its eyes, however, still continued to be fixed upon her, she said after the lapse of a few minutes, " Why don't you speak —— ?" The figure immediately moved off towards the window at the farther end of the room, with its eyes still gazing on her, and it passed so very close to her in doing so, that she was struck by the circumstance of hearing no step nor sound, nor feeling her clothes brushed against, nor even any agitation in the air. Although she was now convinced that the figure was

not her husband, yet she never for a moment supposed that it was any thing supernatural, and was
soon convinced that it was a spectral illusion.   As
soon as this conviction had established itself in her
mind, she recollected the experiment which I had
suggested, of trying to double the object ; but before
she was able distinctly to do this, the figure had retreated to the window, where it disappeared.   Mrs
A. immediately followed it, shook the curtains and
examined the window, the impression having been
so distinct and forcible that she was unwilling to
believe that it was not a reality.   Finding, however,
that the figure had no natural means of escape, she
was convinced that she had seen a spectral apparition
like those recorded in Dr Hibbert's work, and she
consequently felt no alarm or agitation.   The appearance was seen in bright day light, and lasted four
or five minutes.   When the figure stood close to
her it concealed the real objects behind it, and the
apparition was fully as vivid as the reality.

3. On these two occasions Mrs A. was alone, but
when the next phantasm appeared her husband was
present.   This took place on the 4th of January
1830.   About ten o'clock at night, when Mr and
Mrs A. were sitting in the drawing-room, Mr A.
took up the poker to stir the fire, and when he was
in the act of doing this, Mrs A. exclaimed, " Why
there's the cat in the room !"—" Where?" asked Mr
A. " There, close to you," she replied.—" Where?" he
repeated.   " Why on the rug to be sure, between yourself and the coal scuttle."   Mr A., who had still the
poker in his hand, pushed it in the direction mentioned; " Take care," cried Mrs A. " take care, you are
hitting her with the poker."   Mr A. again asked her

to point out exactly where she saw the cat. She replied, "Why sitting up there close to your feet on the rug : She is looking at me. It is Kitty—come here Kitty?"—There were two cats in the house, one of which went by this name, and they were rarely if ever in the drawing-room. At this time Mrs A. had no idea that the sight of the cat was an illusion. When she was asked to touch it, she got up for the purpose, and seemed as if she were pursuing something which moved away. She followed a few steps, and then said, "It has gone under the chair." Mr A. assured her it was an illusion, but she would not believe it. He then lifted up the chair, and Mrs A. saw nothing more of it. The room was then searched all over, and nothing found in it. There was a dog lying on the hearth, who would have betrayed great uneasiness if a cat had been in the room, but he lay perfectly quiet. In order to be quite certain, Mr A. rung the bell, and sent for the two cats, both of which were found in the housekeeper's room.

4. About a month after this occurrence, Mrs A., who had taken a somewhat fatiguing drive during the day, was preparing to go to bed about eleven o'clock at night, and, sitting before the dressing-glass, was occupied in arranging her hair. She was in a listless and drowsy state of mind, but fully awake. When her fingers were in active motion among the papillotes, she was suddenly startled by seeing in the mirror the figure of a near relation, who was then in Scotland, and in perfect health. The apparition appeared over her left shoulder, and its eyes met her's in the glass. It was enveloped in grave-clothes, closely pinned, as is usual with corpses, round the head, and under the chin, and

though the eyes were open, the features were solemn and rigid. The dress was evidently a shroud, as Mrs A. remarked even the punctured pattern usually worked in a peculiar manner round the edges of that garment. Mrs A. described herself as at the time sensible of a feeling like what we conceive of fascination, compelling her for a time to gaze on this melancholy apparition, which was as distinct and vivid as any reflected reality could be, the light of the candles upon the dressing-table appearing to shine fully upon its face. After a few minutes, she turned round to look for the reality of the form over her shoulder; but it was not visible, and it had also disappeared from the glass when she looked again in that direction.

5. In the beginning of March, when Mr A. had been about a fortnight from home, Mrs A. frequently heard him moving near her. Nearly every night as she lay awake, she distinctly heard sounds like his breathing hard on the pillow by her side, and other sounds such as he might make while turning in bed.

6. On another occasion, during Mr A.'s absence, while riding with a neighbour Mr ——, she heard his voice frequently as if he were riding by his side. She heard also the tramp of his horse's feet, and was almost puzzled by hearing him address her at the same time with the person really in company. His voice made remarks on the scenery, improvements, &c. such as he probably should have done had he been present. On this occasion, however, there was no visible apparition.

7. On the 17th March, Mrs A. was preparing for bed. She had dismissed her maid, and was sit-

ting with her feet in hot water. Having an excellent memory, she had been thinking upon and repeating to herself a striking passage in the Edinburgh Review, when, on raising her eyes, she saw seated in a large easy chair before her the figure of a deceased friend, the sister of Mr A. The figure was dressed, as had been usual with her, with great neatness, but in a gown of a peculiar kind, such as Mrs A. had never seen her wear, but exactly such as had been described to her by a common friend as having been worn by Mr A.'s sister during her last visit to England. Mrs A. paid particular attention to the dress, air, and appearance of the figure, which sat in an easy attitude in the chair, holding a handkerchief in one hand. Mrs A. tried to speak to it, but experienced a difficulty in doing so, and in about three minutes the figure disappeared. About a minute afterwards, Mr A. came into the room, and found Mrs A. slightly nervous, but fully aware of the delusive nature of the apparition. She described it as having all the vivid colouring and apparent reality of life ; and for some hours preceding this and other visions, she experienced a peculiar sensation in her eyes, which seemed to be relieved when the vision had ceased.

8. On the 5th October, between one and two o'clock in the morning, Mr A. was awoke by Mrs A., who told him that she had just seen the figure of his deceased mother draw aside the bed-curtains and appear between them. The dress and the look of the apparition were precisely those in which Mr A.'s mother had been last seen by Mrs A. at Paris in 1824.

9. On the 11th October, when sitting in the

drawing-room, on one side of the fire-place, she saw the figure of another deceased friend moving towards her from the window at the farther end of the room. It approached the fire-place, and sat down in the chair opposite. As there were several persons in the room at the time, she describes the idea uppermost in her mind to have been a fear lest they should be alarmed at her staring, in the way she was conscious of doing, at vacancy, and should fancy her intellect disordered. Under the influence of this fear, and recollecting a story of a similar effect in your work on Demonology, which she had lately read, she summoned up the requisite resolution to enable her to cross the space before the fire-place, and seat herself in the same chair with the figure. The apparition remained perfectly distinct till she sat down, as it were, in its lap, when it vanished.

10. On the 26th of the same month, about two P. M. Mrs A. was sitting in a chair by the window in the same room with her husband. He heard her exclaim, " What have I seen ?" And on looking at her, he observed a strange expression in her eyes and countenance. A carriage and four had appeared to her to be driving up the entrance road to the house. As it approached, she felt inclined to go up stairs to prepare to receive company, but, as if spell-bound, she was unable to move or speak. The carriage approached, and as it arrived within a few yards of the window, she saw the figures of the postillions and the persons inside take the ghastly appearance of skeletons and other hideous figures. The whole then vanished entirely, when she uttered the above-mentioned exclamation.

11. On the morning of the 30th October, when Mrs A. was sitting in her own room with a favourite dog in her lap, she distinctly saw the same dog moving about the room during the space of about a minute or rather more.

12. On the 3d December, about nine P. M. when Mr and Mrs A. were sitting near each other in the drawing-room occupied in reading, Mr A. felt a pressure on his foot. On looking up, he observed Mrs A.'s eyes fixed with a strong and unnatural stare on a chair about nine or ten feet distant. Upon asking her what she saw, the expression of her countenance changed, and upon recovering herself, she told Mr A. that she had seen his brother, who was alive and well at the moment in London, seated in the opposite chair, but dressed in grave-clothes, and with a ghastly countenance, as if scarcely alive.

Such is a brief account of the various spectral illusions observed by Mrs A.—In describing them I have used the very words employed by her husband in his communications to me on the subject; * and the reader may be assured that the descriptions are neither heightened by fancy, nor amplified by invention. The high character and intelligence of the lady, and the station of her husband in society, and as a man of learning and science, would authenticate the most marvellous narrative, and satisfy the most scrupulous mind, that the case has been philosophically as well as faithfully described. In narrating events which we regard as of a supernatural character, the mind has a strong tendency to give

* *Edinburgh Journal of Science,* New Series, No. iv. p. 218, 219. No. vi. p. 244, and No. viii. p. 261.

more prominence to what appears to itself the most wonderful ; but from the very same cause, when we describe extraordinary and inexplicable phenomena which we believe to be the result of natural causes, the mind is prone to strip them of their most marvellous points, and bring them down to the level of ordinary events. From the very commencement of the spectral illusions seen by Mrs A. both she and her husband were well aware of their nature and origin, and both of them paid the most minute attention to the circumstances which accompanied them, not only with the view of throwing light upon so curious a subject, but for the purpose of ascertaining their connection with the state of health under which they appeared.

As the spectres seen by Nicolai and others had their origin in bodily indisposition, it becomes interesting to learn the state of Mrs A.'s health when she was under the influence of these illusions. During the six weeks within which the three first illusions took place, she had been considerably reduced and weakened by a troublesome cough, and the weakness which this occasioned was increased by her being prevented from taking a daily tonic. Her general health had not been strong, and long experience has put it beyond a doubt, that her indisposition arises from a disordered state of the digestive organs. Mrs A. has naturally a morbidly sensitive imagination, which so painfully affects her corporeal impressions, that the account of any person having suffered severe pain by accident or otherwise occasionally produces acute twinges of pain in the corresponding parts of her person. The account, for example, of the amputation of an arm will produce

an instantaneous and severe sense of pain in her own
arm.   She is subject to talk in her sleep with great
fluency, to repeat long passages of poetry, particularly when she is unwell, and even to cap verses
for half an hour together, never failing to quote
lines beginning with the final letter of the preceding one till her memory is exhausted.

Although it is not probable that we shall ever
be able to understand the actual manner in which
a person of sound mind beholds spectral apparitions
in the broad light of day, yet we may arrive at such
a degree of knowledge on the subject as to satisfy
rational curiosity, and to strip the phenomena of
every attribute of the marvellous.   Even the vision
of natural objects presents to us insurmountable
difficulties, if we seek to understand the precise part
which the mind performs in perceiving them ; but
the philosopher considers that he has given a satisfactory explanation of vision when he demonstrates
that distinct pictures of external objects are painted on the retina, and that this membrane communicates with the brain by means of nerves of the
same substance as itself, and of which it is merely
an expansion.   Here we reach the gulf which human intelligence cannot pass ; and if the presumptuous mind of man shall dare to extend its speculations farther, it will do it only to evince its in-
capacity and mortify its pride.

In his admirable work on this subject, Dr Hibbert has shown that spectral apparitions are nothing
more than ideas or the recollected images of the
mind, which in certain states of bodily indisposition
have been rendered more vivid than actual impressions, or, to use other words, that the pictures in

the " mind's eye" are more vivid than the pictures in
the body's eye. This principle has been placed by
Dr Hibbert beyond the reach of doubt ; but I propose to go much farther, and to show that the
" mind's eye" is actually the body's eye, and that
the retina is the common tablet on which both
classes of impressions are painted, and by means of
which they receive their visual existence according
to the same optical laws. Nor is this true merely
in the case of spectral illusions : It holds good of
all ideas recalled by the memory or created by the
imagination, and may be regarded as a fundamental
law in the science of pneumatology.

It would be out of place in a work like this to
adduce the experimental evidence on which it rests,
or even to explain the manner in which the experiments themselves must be conducted ; but I may
state in general, that the spectres conjured up by
the memory or the fancy have always a " local ha-
bitation," and that they appear in front of the eye,
and partake in its movements exactly like the impressions of luminous objects after the objects themselves are withdrawn.

In the healthy state of the mind and body, the relative intensity of these two classes of impressions on
the retina are nicely adjusted. The mental pictures
are transient and comparatively feeble, and in ordinary temperaments are never capable of disturbing
or effacing the direct images of visible objects. The
affairs of life could not be carried on if the memory
were to intrude bright representations of the past
into the domestic scene, or scatter them over the
external landscape. The two opposite impressions,
indeed, could not co-exist : The same nervous fibre

D

which is carrying from the brain to the retina the figures of memory, could not at the same instant be carrying back the impressions of external objects from the retina to the brain. The mind cannot perform two different functions at the same instant, and the direction of its attention to one of the two classes of impressions necessarily produce the extinction of the other : But so rapid is the exercise of mental power, that the alternate appearance and disappearance of the two contending impressions is no more recognized than the successive observations of external objects during the twinkling of the eyelids. If we look, for example, at the Façade of St Paul's, and, without changing our position, call to mind the celebrated view of Mont Blanc from Lyons, the picture of the cathedral, though actually impressed upon the retina, is momentarily lost sight of by the mind, exactly like an object seen by indirect vision; and during the instant the recollected image of the mountain, towering over the subjacent range, is distinctly seen, but in a tone of subdued colouring, and indistinct outline. When the purpose of its recall is answered, it quickly disappears, and the picture of the cathedral again resumes the ascendancy.

In darkness and solitude, when external objects no longer interfere with the pictures of the mind, they become more vivid and distinct ; and in the state between waking and sleeping, the intensity of the impressions approaches to that of visible objects. With persons of studious habits, who are much occupied with the operations of their own minds, the mental pictures are much more distinct than in ordinary persons ; and in the midst of ab-

stract thought, external objects even cease to make
any impression on the retina. A philosopher absorbed in his contemplations experiences a temporary privation of the use of his senses. His children or his servants will enter the room directly
before his eyes without being seen. They will
speak to him without being heard; and they will
even try to rouse him from his reverie without being felt; although his eyes, his ears, and his nerves,
actually receive the impressions of light, sound, and
touch. In such cases, however, the philosopher is
voluntarily pursuing a train of thought on which his
mind is deeply interested; but even ordinary men,
not much addicted to speculations of any kind, often perceive in their mind's eye the pictures of deceased or absent friends, or even ludicrous creations
of fancy, which have no connection whatever with
the train of their thoughts. Like spectral apparitions they are entirely involuntary, and though they
may have sprung from a regular series of associations, yet it is frequently impossible to discover a
single link in the chain.

If it be true, then, that the pictures of the mind
and spectral illusions are equally impressions upon
the retina, the latter will differ in no respect from
the former, but in the degree of vividness with
which they are seen; and those frightful apparitions
become nothing more than our ordinary ideas, rendered more brilliant by some accidental and temporary derangement of the vital functions. Their
very vividness too, which is their only characteris-
tic, is capable of explanation. I have already shown
that the retina is rendered more sensible to light by
voluntary local pressure, as well as by the involun-

tary pressure of the blood-vessels behind it ; and if, by looking at the sun, we impress upon the retina a coloured image of that luminary, which is seen even when the eye is shut, we may by pressure alter the colour of that image, in consequence of having increased the sensibility of that part of the retina on which it is impressed.   Hence we may readily understand how the vividness of the mental pictures must be increased by analogous causes.

In the case both of Nicolai and Mrs A. the immediate cause of the spectres was a deranged action of the stomach.   When such a derangement is induced by poison, or by substances which act as poisons, the retina is peculiarly affected, and the phenomena of vision singularly changed.   Dr Patouillet has described the case of a family of *nine* persons who were all driven mad by eating the root of the *Hyoscyamus niger* or black Henbane.   One of them leapt into a pond.   Another exclaimed that his neighbour would lose a cow in a month, and a third vociferated that the crown piece of sixty pence would in a short time rise to five livres.   On the following day they had all recovered their senses, but recollected nothing of what had happened.   On the same day they all saw objects double, and, what is still more remarkable, on the third day *every object appeared to them as red as scarlet*.   Now this red light was probably nothing more than the red phosphorescence produced by the pressure of the blood-vessels on the retina, and analogous to the masses of *blue, green, yellow,* and *red* light, which have been already mentioned as produced by a similar pressure in headaches, arising from a disordered state of the digestive organs.

Were we to analyse the various phenomena of spectral illusions, we should discover many circumstances favourable to these views. In those seen by Nicolai the individual figures were always somewhat paler than natural objects. They sometimes grew more and more indistinct, and became perfectly white ; and, to use his own words, "he could always distinguish with the greatest precision, phantasms from phenomena." Nicolai sometimes saw the spectres when his eyes were shut, and sometimes they were thus made to disappear,—effects perfectly identical with those which arise from the impressions of very luminous objects. Sometimes the figures vanished entirely, and at other times only pieces of them disappeared, exactly conformable to what takes place with objects seen by indirect vision, which most of those figures must necessarily have been.

Among the peculiarities of spectral illusions there is one which merits particular attention, namely, that they seem to cover or conceal objects immediately beyond them. It is this circumstance more than any other which gives them the character of reality, and at first sight it seems difficult of explanation. The distinctness of any impression on the retina is entirely independent of the accommodation of the eye to the distinct vision of external objects. When the eye is at rest, and is not accommodated to objects at any particular distance, it is in a state for seeing distant objects most perfectly. When a distinct spectral impression, therefore, is before it, all other objects in its vicinity will be seen indistinctly, for while the eye is engrossed with the vision, it is not likely to accommodate itself to any other object in the same direction. It is quite com-

mon, too, for the eye to see only one of two objects actually presented to it. A sportsman who
has been in the practice of shooting with both his
eyes open actually sees a double image of the muzzle of his fowling-piece, though it is only with one
of these images that he covers his game, having no
perception whatever of the other. But there is still
another principle upon which only one of two
objects may be seen at a time. If we look very
steadily and continuously at a double pattern, such
as those on a carpet composed of two single patterns
of different colours, suppose *red* and *yellow* ; and if
we direct the mind particularly to the contemplation of the red one, the green pattern will sometimes
vanish entirely, leaving the red one alone visible,
and by the same process the *red* one may be made
to disappear. In this case, however, the two patterns, like the two images, may be seen together ;
but if the very same portion of the retina is excited
by the direct rays of an external object, when it is
excited by a mental impression, it can no more see
them both at the same time than a vibrating string
can give out two different fundamental sounds. It
is quite possible, however, that the brightest parts of
a spectral figure may be distinctly seen along with
the brightest parts of an object immediately behind
it, but then the bright parts of each object will fall
upon different parts of the retina.

These views are illustrated by a case mentioned by
Dr Abercrombie. A gentleman, who was a patient
of his, of an irritable habit, and liable to a variety of
uneasy sensations in his head, was sitting alone in
his dining-room in the twilight, when the door
of the room was a little open. He saw distinctly

a female figure enter, wrapped in a mantle, with the face concealed by a large black bonnet. She seemed to advance a few steps towards him, and then stop. He had a full conviction that the figure was an illusion of vision, and he amused himself for some time by watching it; at the same time observing that he could see through the figure so as to perceive the lock of the door, and other objects behind it. *

If these views be correct the phenomena of spectral apparitions are stripped of all their terror, whether we view them in their supernatural character or as indications of bodily indisposition. Nicolai, even, in whose case they were accompanied with alarming symptoms, derived pleasure from the contemplation of them, and he not only recovered from the complaint in which they originated, but survived them for many years.—Mrs A., too, who sees them only at distant intervals, and with whom they have but a fleeting existence, will, we trust, soon lose her exclusive privilege, when the slight indisposition which gives them birth has subsided.

* Inquiries concerning the Intellectual Powers, and the Investigation of Truth. Edinburgh, 1830.

LETTER IV.

*Science used as an instrument of imposture—Deceptions with
plane and concave mirrors practised by the ancients—The
magician's mirror — Effects of concave mirrors —Aërial
Images—Images on smoke—Combination of mirrors for producing pictures from living objects—The mysterious dagger
—Ancient miracles with concave mirrors—Modern necromancy with them, as seen by Cellini—Description and effects
of the magic lantern—Improvements upon it—Phantasma-
goric exhibitions of Philipstal and others—Dr Young's arrangement of Lenses, &c. for the Phantasmagoria—Improvements suggested—Catadioptrical phantasmagoria for producing the pictures from living objects—Method of cutting
off parts of the figures—Kircher's mysterious handwriting
on the wall—His hollow cylindrical mirror for aërial
images—Cylindrical mirror for reforming distorted pictures
—Mirrors of variable curvature for producing caricatures.*

In the preceding observations man appears as
the victim of his own delusions—as the magician
unable to exorcise the spirits which he has himself
called into being. We shall now see him the dupe
of preconcerted imposture—the slave of his own
ignorance—the prostrate vassal of power and superstition. I have already stated that the monarchs
and priests of ancient times carried on a systematic
plan of imposing upon their subjects—a mode of government which was in perfect accordance with

their religious belief: But it will scarcely be believed that the same delusions were practised after the establishment of Christianity, and that even the Catholic sanctuary was often the seat of these unhallowed machinations.   Nor was it merely the low and cunning priest who thus sought to extort money and respect from the most ignorant of his flock: Bishops and pontiffs themselves wielded the magician's wand over the diadems of kings and emperors, and, by the pretended exhibition of supernatural power, made the mightiest potentates of Europe tremble upon their thrones.   It was the light of science alone which dispelled this moral and intellectual darkness, and it is entirely in consequence of its wide diffusion that we live in times when sovereigns seek to reign only through the affections of their people, and when the minister of religion asks no other reverence but that which is inspired by the sanctity of his office and the purity of his character.

It was fortunate for the human race that the scanty knowledge of former ages afforded so few elements of deception.   What a tremendous engine would have been worked against our species by the varied and powerful machinery of modern science ! Man would still have worn the shackles which it forged, and his noble spirit would still have groaned beneath its fatal pressure.

There can be little doubt that the most common, as well as the most successful, impositions of the ancients were of an optical nature, and were practised by means of plane and concave mirrors.   It has been clearly shown by various writers that the ancients made use of mirrors of steel, silver, and a composition of copper and tin, like those now used for

reflecting specula. It is also very probable from a
passage in Pliny, that glass mirrors were made at
Sidon ; but it is evident, that, unless the object presented to them was illuminated in a very high degree, the images which they formed must have been
very faint and unsatisfactory. The silver mirrors,
therefore, which were universally used, and which
are superior to those made of any other metal,
are likely to have been most generally employed by
the ancient magicians. They were made to give
multiplied and inverted images of objects, that is,
they were plane, polygonal or many-sided, and
concave. There is one property, however, mentioned by Aulus Gellius, which has given unnecessary
perplexity to commentators. He states that there
were specula, which, when put in a particular place,
gave no images of objects, but, when carried to another place, recovered their property of reflexion. *
M. Salverte is of opinion that, in quoting Varro,
Aulus Gellius was not sufficiently acquainted with
the subject, and erred in supposing that the phenomenon depended on the *place* instead of the position of the mirror ; but this criticism is obviously made with the view of supporting an opinion of
his own, that the property in question may be analogous to the phenomenon of polarised light, which
at a certain angle refuses to suffer reflexion from
particular bodies. If this idea has any foundation,
the mirror must have been of glass or some other
body not metallic, or, to speak more correctly,
there must have been *two* such mirrors, so nice-

* *Ut speculum in loco certo positum nihil imaginet ; alior-
sum translatum faciat imagines.*   Aul. Gel. Noct. Attic. lib.
xvi. cap. 18.

ly adjusted not only to one another, but to the light incident upon each, that the effect could not possibly be produced but by a philosopher thoroughly acquainted with the modern discovery of the polarisation of light by reflexion. Without seeking for so profound an explanation of the phenomenon, we may readily understand how a silver mirror may instantly lose its reflecting power, in a damp atmosphere, in consequence of the precipitation of moisture upon its surface, and may immediately recover it when transported into drier air.

One of the simplest instruments of optical deception is the plane mirror, and when two are combined for this purpose it has been called the magician's mirror. An observer in front of a plane mirror sees a distinct image of himself; but if two persons take up a mirror, and if the one person is as much to one side of a line perpendicular to the middle of it as the other is to the other side, they will

Fig. 3.

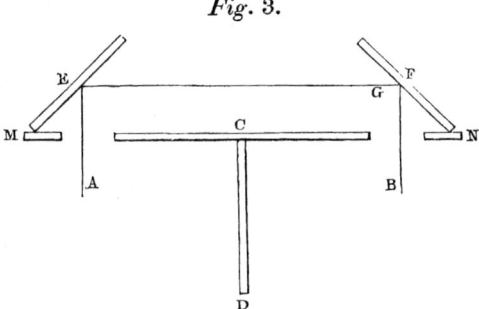

see each other but not themselves. If we now suppose MC, CD, NC, CD to be the partitions of two

adjacent apartments, let square openings be made
in the partitions at A and B, about five feet above
the floor, and let them be filled with plate glass, and
surrounded with a picture frame, so as to have the
appearance of two mirrors.   Place two mirrors E,
F, one behind each opening at A and B, inclined
45° to the partition MN, and so large that a person
looking into the plates of glass at A and B will not see
their edges.   When this is done it is obvious that
a person looking into the mirror A will not see himself but will see any person or figure placed at B.
If he believes that he is looking into a common
mirror at A, his astonishment will be great at seeing himself transformed into another person, or into any living animal that may be placed at B.   The
success of this deception would be greatly increased
if a plane mirror suspended by a pulley could be
brought immediately behind the plane glass at A,
and drawn up from it at pleasure.   The spectator
at A having previously seen himself in this moveable mirror, would be still more astonished when
he afterwards perceived in the same place a face different from his own.   By drawing the moveable mirror half up, the spectator at A might see half of his
own face joined to half of the face placed at B ;
but in the present day the most ignorant persons
are so familiar with the properties of a looking glass
that it would be very difficult to employ this kind of
deception with the same success which must have
attended it in a more illiterate age.   The optical
reader will easily see that the mirror F and the apartment NCD are not absolutely necessary for carrying on this deception ; for the very same effects
will be produced if the person at B is stationed at

G, and looks towards the mirror F in the direction GF. As the mirror F, however, must be placed as near to A as possible, the person at G would be too near the partition C N, unless the mirror F was extremely large.

The effect of this and every similar deception is greatly increased when the persons are illuminated with a strong light, and the rest of the apartment as dark as possible ; but whatever precautions are taken, and however skilfully plane mirrors are combined, it is not easy to produce with them any very successful illusions.

The concave mirror is the staple instrument of the magician's cabinet, and must always perform a principal part in all optical combinations. In order to be quite perfect, every concave mirror should have its surface elliptical, so that if any object is placed in one focus of the ellipse, an inverted image of it will be formed in the other focus. This image, to a spectator rightly placed, appears suspended in the air, so that if the mirror and the object are hid from his view, the effect must appear to him almost supernatural.

The method of exhibiting the effect of concave mirrors most advantageously is shown in Fig. 3, where CD is the partition of a room having in it a square opening EF, the centre of which is about five feet above the floor. This opening might be surrounded with a picture frame, and a painting which exactly filled it might be so connected with a pulley that it could be either slipped aside, or raised so as to leave the frame empty. A large concave mirror MN is then placed in another apartment, so that when any object is placed at A, a distinct image of

it may be formed in the centre of the opening EF.
Let us suppose this object to be a plaster cast of

Fig. 4.

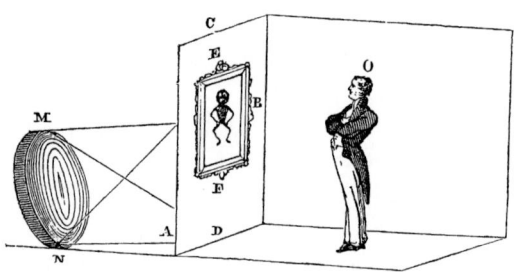

any object made as white as possible, and placed
in an *inverted* position at A.   A strong light
should then be thrown upon it by a powerful
lamp, the rays of which are prevented from reaching the opening EF.   When this is done, a spectator placed at O will see an erect image of the statue at B the centre of the opening—standing in the
air, and differing from the real statue only in being
a little larger, while the apparition will be wholly
invisible to other spectators placed at a little distance on each side of him.

If the opening EF is filled with smoke rising
either from a chafing dish, in which incense is burnt,
or made to issue in clouds from some opening below, the image will appear in the middle of the
smoke depicted upon it as upon a ground, and capable of being seen by those spectators who could
not see the image in the air.   The rays of light, in
place of proceeding without obstruction to an eye

at O, are reflected as it were from those minute particles of which the smoke is composed, in the same manner as a beam of light is rendered more visible by passing through an apartment filled with dust or smoke.

It has long been a favourite experiment to place at A, a white and strongly illuminated human skull, and to exhibit an image of it amid the smoke of a chafing dish at B ; but a more terrific effect would be produced if a small skeleton, suspended by invisible wires, were placed as an object at A. Its image suspended in the air at B, or painted upon smoke, could not fail to astonish the spectator.

The difficulty of placing a living person in an inverted position, as an object at A, has no doubt prevented the optical conjuror from availing himself of so admirable a resource ; but this difficulty may be removed by employing a second concave mirror. This second mirror must be so placed as to reflect towards MN, the rays proceeding from an erect living object, and to form an inverted image of this object at A. An erect image of this inverted image will then be formed at B, either suspended in the air or depicted upon a wreath of smoke. This aërial image will exhibit the precise form and colours and movements of the living object, and it will maintain its character as an apparition if any attempt is made by the spectator to grasp its unsubstantial fabric.

A deception of an alarming kind, called the *Mysterious dagger*, has been long a favourite exhibition. If a person with a drawn and highly polished dagger, illuminated by a strong light, stands a little farther from a concave mirror than its principal focus, he will perceive in the air between himself and the mirror, an inverted and diminished image of his own person with the dagger similarly brandished : If he aims the dagger at the centre of the mirror's concavity, the two daggers will meet point to point, and, by pushing it still farther from him towards the mirror, the imaginary dagger will strike at his heart. In this case it is necessary that the direction of the real dagger coincides with a diameter of the sphere of which the mirror is a part ; but if its direction is on one side of that diameter, the direction of the imaginary dagger will be as far on the other side of the diameter, and the latter will aim a blow at any person who is placed in the proper position for receiving it. If the person who bears the real dagger is therefore placed behind a screen, or otherwise concealed from the view of the spectator who is made to approach to the place of the image, the thrust of the polished steel at his breast will not fail to produce a powerful impression. The effect of this experiment would no doubt be increased by covering with black cloth the person who holds the dagger, so that the image of his hand only should be seen, as the inverted picture of him would take away from the reality of the appearance. By using two mirrors, indeed, this defect might be remedied, and the spectator would witness an exact image of the assassin aiming the dagger at his life.

The common way of making this experiment is to place a basket of fruit above the dagger, so that a distinct aerial image of the fruit is formed in the focus of the mirror. The spectator having been desired to take some fruit from the basket approaches for that purpose, while a person properly concealed withdraws the real basket of fruit with one hand, and with the other advances the dagger, the image of which, being no longer covered by the fruit, strikes at the body of the astonished spectator.

The powers of the concave mirror have been likewise displayed in exhibiting the apparition of an absent or deceased friend. For this purpose a strongly illuminated bust or picture of the person is placed before the concave mirror, and a distinct image of the picture will be seen either in the air or among smoke in the manner already described. If the back ground of the picture is temporarily covered with lamp-black, so that there is no light about the picture but what falls upon the figure, the effect will be more complete.

As in all experiments with concave mirrors, the size of the aërial image is to that of the real object as their distances from the mirror, we may, by varying the distance of the object, increase or diminish the size of the image. In doing this, however, the distance of the image from the mirror is at the same time changed, so that it would quit the place most suitable for its exhibition. This defect may be removed by simultaneously changing the place both of the mirror and the object, so that the image may remain stationary, expanding itself from a luminous spot to a gigantic size, and again passing through all intermediate magnitudes, till it vanishes in a cloud of light.

Those who have studied the effects of concave mirrors of a small size, and without the precautions necessary to insure deception, cannot form any idea

E

of the magical effect produced by this class of optical apparitions. When the instruments of illusion are themselves concealed,—when all extraneous lights but those which illuminate the real object are excluded,—when the mirrors are large and well polished and truly formed,—the effect of the representation on ignorant minds is altogether overpowering, while even those who know the deception, and perfectly understand its principles, are not a little surprised at its effects. The inferiority in the effects of a common concave mirror to that of a well arranged exhibition, is greater even than that of a perspective picture hanging in an apartment, to the same picture exhibited under all the imposing accompaniments of a dioramic representation.

It can scarcely be doubted, that a concave mirror was the principal instrument by which the heathen gods were made to appear in the ancient temples. In the imperfect accounts which have reached us of these apparitions, we can trace all the elements of an optical illusion. In the ancient temple of Hercules at Tyre, Pliny mentions that there was a seat made of a consecrated stone, "from which the gods easily rose." Esculapius often exhibited himself to his worshippers in his temple at Tarsus; and the temple of Enguinum in Sicily was celebrated as the place where the goddesses exhibited themselves to mortals. Jamblichus actually informs us, that the ancient magicians caused the gods to appear among the vapours disengaged from fire; and when the conjuror Maximus terrified his audience by making the statue of Hecate laugh, while in the middle of the smoke of burning incense, he was obviously dealing with the image of a living object dressed in the costume of the sorceress.

The character of these exhibitions in the ancient temples is so admirably depicted in the following passage of Damascius quoted by M. Salverte, that we recognize all the optical effects which have been already described. " In a manifestation," says he, " which ought not to be revealed ...... there appeared on the wall of the temple a mass of light which at first seemed to be very remote ; it transformed itself in coming nearer, into a face evidently divine and supernatural, of a severe aspect, but mixed with gentleness, and extremely beautiful. According to the institutions of a mysterious religion the Alexandrians honoured it as Osiris and Adonis."

Among more modern examples of this illusion, we may mention the case of the Emperor Basil of Macedonia. Inconsolable at the loss of his son, this sovereign had recourse to the prayers of the Pontiff Theodore Santabaren, who was celebrated for his power of working miracles. The ecclesiastical conjuror exhibited to him the image of his beloved son magnificently dressed and mounted upon a superb charger : The youth rushed towards his father, threw himself into his arms and disappeared. M. Salverte judiciously observes, that this deception could not have been performed by a real person, who imitated the figure of the young prince. The existence of this person, betrayed by so remarkable a resemblance, and by the trick of the exhibition, could not fail to have been discovered and denounced, even if we could explain how the son could be so instantaneously disentangled from his father's embrace. The emperor, in short, saw the aërial image of a picture of his son on horseback, and as the picture was brought nearer the mirror, the image advanced into his arms, when it of course eluded his affectionate grasp.

These and other allusions to the operations of the ancient magic, though sufficiently indicative of the methods which were employed, are too meagre to convey any idea of the splendid and imposing exhibitions which must have been displayed. A national system of deception, intended as an instrument of government, must have brought into requisition not merely the scientific skill of the age, but a variety of subsidiary contrivances, calculated to astonish the beholder, to confound his judgment, to dazzle his senses, and to give a predominant influence to the peculiar imposture which it was thought desirable to establish. The grandeur of the means may be inferred from their efficacy, and from the extent of their influence.

This defect, however, is to a certain degree supplied by an account of a modern necromancy, which has been left us by the celebrated Benvenuto Cellini, and in which he himself performed an active part.

" It happened," says he, " through a variety of odd accidents, that I made acquaintance with a Sicilian priest, who was a man of genius, and well versed in the Latin and Greek authors. Happening one day to have some conversation with him when the subject turned upon the art of necromancy, I, who had a great desire to know something of the matter, told him, that I had all

my life felt a curiosity to be acquainted with the
mysteries of this art.

"The priest made answer, ' That the man must
be of a resolute and steady temper who enters upon
that study.' I replied, ' that I had fortitude and resolution enough, if I could but find an opportunity.'
The priest subjoined, ' if you think you have the
heart to venture, I will give you all the satisfaction you can desire.' Thus we agreed to enter upon a plan of necromancy. The priest one evening
prepared to satisfy me, and desired me to look out
for a companion or two. I invited one Vincenzio
Romoli, who was my intimate acquaintance : he
brought with him a native of Pistoia, who cultivat-
ed the black art himself. We repaired to the
Colosseo, and the priest, according to the custom
of necromancers, began to draw circles upon the
ground, with the most impressive ceremonies ima-
ginable : he likewise brought hither assafœtida,
several precious perfumes, and fire, with some compositions also, which diffused noisome odours.   As
soon as he was in readiness, he made an opening
to the circle, and having taken us by the hand, or-
dered the other necromancer, his partner, to throw
the perfumes into the fire at a proper time, intrust-
ing the care of the fire and perfumes to the rest ;
and thus he began his incantations. This ceremony lasted above an hour and a-half, when there appeared several legions of devils, insomuch that the
amphitheatre was quite filled with them. I was
busy about the perfumes, when the priest, perceiving there was a considerable number of infernal
spirits, turned to me and said, ' Benvenuto, ask
them something?' I answered, ' let them bring me

into the company of my Sicilian mistress Angelica.' That night he obtained no answer of any
sort ; but I had received great satisfaction in having my curiosity so far indulged. The necromancer told me it was requisite we should go a second
time, assuring me that I should be satisfied in
whatever I asked ; but that I must bring with me
a pure immaculate boy.

" I took with me a youth who was in my service,
of about twelve years of age, together with the
same Vincenzio Romoli, who had been my companion the first time, and one Agnolino Gaddi, an intimate acquaintance, whom I likewise prevailed on
to assist at the ceremony. When we came to the
place appointed, the priest having made his prepa-
rations as before, with the same and even more
striking ceremonies, placed us within the circle,
which he had likewise drawn with a more wonderful art, and in a more solemn manner than at our
former meeting. Thus, having committed the care
of the perfumes and the fire to my friend Vincenzio, who was assisted by Agnolino Gaddi, he put
into my hand a pintaculo or magical chart, and bid
me turn it towards the places that he should direct
me ; and under the pintaculo I held the boy. The
necromancer, having begun to make his tremendous
invocations, called by their names a multitude of
demons who were the leaders of the several legions,
and questioned them, by the power of the eternal
uncreated God, who lives for ever, in the Hebrew
language, as likewise in Latin and Greek ; insomuch
that the amphitheatre was almost in an instant filled with demons more numerous than at the former conjuration. Vincenzio Romoli was busied in

making a fire, with the assistance of Agnolino, and burning a great quantity of precious perfumes. I, by the directions of the necromancer, again desired to be in the company of my Angelica. The former thereupon turning to me, said,—' Know, they have declared that in the space of a month you shall be in her company.'

" He thus requested me to stand resolutely by him, because the legions were now above a thousand more in number than he had designed ; and besides, these were the most dangerous ; so that, after they had answered my question, it behoved him to be civil to them and dismiss them quietly. At the same time the boy under the pintaculo was in a terrible fright, saying, that there were in that place a million of fierce men, who threatened to destroy us ; and that, moreover, four armed giants of enormous stature were endeavouring to break into our circle. During this time, whilst the necromancer, trembling with fear, endeavoured by mild and gentle methods to dismiss them in the best way he could, Vincenzio Romoli, who quivered like an aspen leaf, took care of the perfumes. Though I was as much terrified as any of them, I did my utmost to conceal the terror I felt ; so that I greatly contributed to inspire the rest with resolution ; but the truth is, I gave myself over for a dead man, seeing the horrid fright the necromancer was in. The boy placed his head between his knees and said, ' In this posture will I die ; for we shall all surely perish,' I told him that all these demons were under us, and what he saw was smoke and shadow ; so bid him hold up his head and take courage. No sooner did he look up than he cried out, ' The whole amphitheatre is burning and the fire is just falling upon us.' So covering his eyes with his hands, he again exclaimed, ' that destruction was inevitable, and desired to see no more.' The necromancer entreated me to have a good heart, and take care to burn proper perfumes; upon which I turned to Romoli, and bid him burn all the most precious perfumes he had. At the same time, I cast my eye upon Agnolino Gaddi, who was terrified to such a degree that he could scarce distinguish objects, and seemed to be half-dead. Seeing him in this condition, I said, ' Agnolino, upon these occasions a man should not yield to fear, but should stir about and give his assistance, so come directly and put on some more of these.' The effects of poor Agnolino's fear were overpowering. The boy hearing a crepitation ventured once more to raise his head, when, seeing me laugh, he began to take courage, and said ' That the devils were flying away with a vengeance.'

" In this condition we stayed till the bell rung for morning prayers. The boy again told us, that there remained but few devils, and these were at a great distance. When the magician had performed the rest of his ceremonies, he stripped off his gown, and took up a wallet full of books which he had brought with him.

" We all went out of the circle together, keeping as close to each other as we possibly could, especially the boy, who had placed himself in the middle, holding the necromancer by the coat, and me by the cloak. As we were going to our houses in the quarter of Banchi, the boy told us that two of the demons whom we had seen at the amphitheatre went on before us leaping and skipping, sometimes

running upon the roofs of the houses, and sometimes upon the ground. The priest declared, that
though he had often entered magic circles, nothing
so extraordinary had ever happened to him. As
we went along, he would fain persuade me to assist
with him at consecrating a brook from which, he
said, we should derive immense riches : we should
then ask the demons to discover to us, the various
treasures with which the earth abounds, which would
raise us to opulence and power ; but that these love-
affairs were mere follies, from whence no good could
be expected. I answered, ' That I would readily
have accepted his proposal, if I understood Latin.'
He redoubled his persuasions, assuring me, that
the knowledge of the Latin language was by no
means material. He added, that he could have
Latin scholars enough, if he had thought it worth
while to look out for them, but that he could never
have met with a partner of resolution and intrepi-
dity equal to mine, and that I should by all means
follow his advice. Whilst we were engaged in this
conversation, we arrived at our respective houses,
and all that night dreamt of nothing but devils."

It is impossible to peruse the preceding description without being satisfied that the legions of devils were not produced by any influence upon the
imaginations of the spectators, but were actual optical phantasms, or the images of pictures or objects produced by one or more concave mirrors or
lenses. A fire is lighted, and perfumes and incense
are burnt, in order to create a ground for the images, and the beholders are rigidly confined within
the pale of the magic circle. The concave mirror
and the objects presented to it having been so

placed that the persons within the circle could not see the aërial image of the objects by the rays directly reflected from the mirror, the work of deception was ready to begin. The attendance of the magician upon his mirror was by no means necessary. He took his place along with the spectors within the magic circle. The images of the devils were all distinctly formed in the air immediately above the fire, but none of them could be seen by those within the circle. The moment, however, that perfumes were thrown into the fire to produce smoke, the first wreath of smoke that rose through the place of one or more of the images, would reflect them to the eyes of the spectator, and they could again disappear if the wreath was not followed by another. More and more images would be rendered visible as new wreaths of smoke arose, and the whole group would appear at once when the smoke was uniformly diffused over the place occupied by the images.

The "compositions which diffused noisome odours" were intended to intoxicate or stupify the spectators, so as to increase their liability to deception, or to add to the real phantasms which were before their eyes others which were the offspring only of their own imaginations. It is not easy to gather from the description what parts of the exhibition were actually presented to the eyes of the spectators, and what parts of it were imagined by themselves. It is quite evident that the boy as well as Agnolino Gaddi, were so overpowered with terror that they fancied many things which they did not see ; but when the boy declares that four armed giants of an enormous stature were threatening to break into their circle, he gives an accurate description of the effect that would be produced by pushing the figures nearer the mirror, and then magnifying their images, and causing them to advance towards the circle.   Although Cellini declares that he was trembling with fear, yet it is quite evident that he was not entirely ignorant of the machinery which was at work, for in order to encourage the boy, who was almost dead with fear, he assured them that the devils were under their power, and that " what he saw was smoke and shadow."

Mr Roscoe, from whose Life of Cellini the preceding description is taken, draws a similar conclusion from the consolatory words addressed to the boy, and states that they " confirm him in the belief, that the whole of these appearances, like a phantasmagoria, were merely the effects of a magic lantern produced on volumes of smoke from various kinds of burning wood."   In drawing this conclusion, Mr Roscoe has not adverted to the fact, that this exhibition took place about the middle of the 16th century, while the magic lantern was not invented by Kircher till towards the middle of the 17th century, Cellini having died in 1570, and Kircher having been born in 1601.   There is no doubt that the effects described could be produced by this instrument, but we are not entitled to have recourse to any other means of explanation but those which were known to exist at the time of Cellini.   If we suppose, however, that the necromancer either had a regular magic lantern, or that he had fitted up his concave mirror in a box containing the figures of his devils, and that this box with its lights was carried home with the party, we can easily account for the declaration of the boy, " that, as they were going home to their houses in the quarter of Banchi, *two of the demons whom we had seen at the Amphitheatre, went on before us leaping and skipping, sometimes running upon the roofs of the houses, and sometimes upon the ground.*"

The introduction of the magic lantern as an optical instrument, supplied the magicians of the 17th century with one of their most valuable tools. The use of the concave mirror, which does not appear to have been even put up into the form of an instrument, required a separate apartment, or at least that degree of concealment which it was difficult on ordinary occasions to command ; but the magic lantern, containing in a small compass its lamp, its lenses, and its sliding figures, was peculiarly fitted for the itinerant conjuror, who had neither the means of providing a less portable and more expensive apparatus, nor the power of transporting and erecting it.

The magic lantern shown in the annexed figure consists of a dark lantern, A B, containing a lamp G, and a concave metallic mirror M N, and it is so constructed that when the lamp is lighted, not a ray of light is able to escape from it. Into the side of the lantern is fitted a double tube C D, the outer half of which D, is capable of moving within the other half. A large plano-convex lens C, is fixed at the inner end of the double tube, and a small convex lens D, at the outer end ; and to the fixed tube C E, there is joined a groove E F, in which the sliders containing the painted objects are placed.

and through which they can be moved. Each slider contains a series of figures or pictures paint-

Fig. 5.

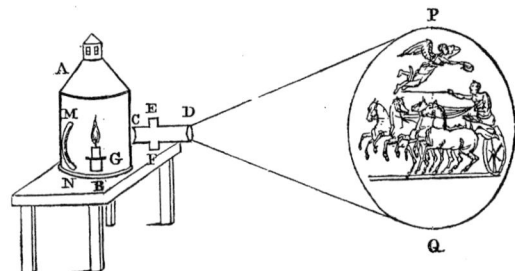

ed on glass with highly transparent colours. The direct light of the lamp G, and the light reflected from the mirror M N, falling upon the illuminating lens C, is concentrated by it so as to throw a brilliant light upon the painting on the slider, and as this painting is in the conjugate focus of the convex lens D, a magnified image of it will be formed on a white wall or white cloth placed at P Q. If the lens D is brought nearer to E F, or to the picture, the distinct image will be more magnified, and will be formed at a greater distance from D, so that if there is any particular distance of the image which is more convenient than another, or any particular size of the object which we wish, it can be obtained by varying the distance of the lens D from E F.

When the image is received on an opaque ground, as is commonly the case, the spectators are placed in the same room with the lantern; but for the

purposes of deception, it would be necessary to place
the lantern in another apartment like the mirror
in Fig 4, and to throw the magnified pictures on a
large plate of ground glass, or a transparent gauze
screen, stretched across an opening E F, Fig. 4,
made in the partition which separates the spectators
from the exhibitor.   The images might, like those
of the concave mirror, be received upon wreaths of
smoke.   These images are of course always inverted in reference to the position of the painted objects ; but in order to render them really erect, we
have only to invert the sliders.   The representations of the magic lantern never fail to excite a
high degree of interest, even when exhibited with
the ordinary apparatus ; but by using double sliders, and varying their movements, very striking
effects may be produced.   A smith, for example,
is made to hammer upon his anvil,—a figure is
thrown into the attitude of terror by the introduction of a spectral apparition,—and a tempest at sea
is imitated, by having the sea on one slider, and the
ships on other sliders, to which an undulatory motion is communicated.

The magic lantern is susceptible of great improvement in the painting of the figures, and in
the mechanism and combination of the sliders.   A
painted figure which appears well executed to the
unassisted eye, becomes a mere daub when magnified 50 or 100 times ; and when we consider what
kind of artists are employed in their execution, we
need not wonder that this optical instrument has
degenerated into a mere toy for the amusement of
the young.   Unless for public exhibition, the expence of exceedingly minute and spirited drawings

could not be afforded; but I have no doubt that if such drawings were executed, a great part of the expence might be saved by engraving them on wood, and transferring their outline to the glass sliders.

A series of curious representations might be effected, by inserting glass plates containing suitable figures in a trough having two of its sides parallel, and made of plate glass. The trough must be introduced at E F, so that the figure on the glass is at the proper distance from the object lens D. When the trough is filled with water or with any transparent fluid, the picture at P Q will be seen with the same distinctness as if the figure had been introduced by itself into the groove E F; but if any transparent fluid of a different density from water is mixed with it, so as to combine with it quickly or slowly, the appearance of the figure displayed at P Q will undergo singular changes. If spirits of wine, or any ardent spirit, are mixed with the water, so as to produce throughout its mass partial variations of density, the figure at P Q will be as it were broken down into a thousand parts, and will recover its continuity and distinctness when the two fluids have combined. If a fluid of less density than water is laid gently upon the water, so as to mix with it gradually, and produce a regular diminution of density downwards ; or if saline substances soluble in water are laid at the bottom of the trough, the density will diminish upwards, and the figure will undergo the most curious elongations and contractions. Analogous effects may be produced by the application of heat to the surface or sides of the trough, so that we may effect at the same time both an increase and a diminution

in the density of the water, in consequence of
which the magnified images will undergo the most
remarkable transformations. It is not necessary
to place the glass plate which contains the figure
within the trough. It may be placed in front of it,
and by thus creating as it were an atmosphere with
local variations of density we may exhibit the phenomena of the mirage and of looming, in which
the inverted images of ships and other objects are
seen in the air, as described in another letter.

The power of the magic lantern has been greatly
extended by placing it on one side of the transparent
screen of taffetas which receives the images while the
spectators are placed on the other side, and by making every part of the glass sliders opaque, excepting
the part which forms the figures.   Hence all the
figures appear luminous on a black ground, and produce a much greater effect with the same degree of
illumination.   An exhibition depending on these
principles was brought out by M. Philipstal in 1802
under the name of the *Phantasmagoria,* and when
it was shown in London and Edinburgh it produced the most impressive effects upon the spectators.   The small theatre of exhibition was lighted
only by one hanging lamp, the flame of which was
drawn up into an opaque chimney or shade when
the performance began.   In this " darkness visible"
the curtain rose and displayed a cave with skeletons and other terrific figures in relief upon its walls.
The flickering light was then drawn up beneath its
shroud, and the spectators in total darkness found
themselves in the middle of thunder and lightning.   A thin transparent screen had, unknown to
the spectators, been let down after the disappear-

ance of the light, and upon it the flashes of lightning and all the subsequent appearances were represented. This screen being half-way between the spectators and the cave which was first shown, and being itself invisible, prevented the observers from having any idea of the real distance of the figures, and gave them the entire character of aërial pictures. The thunder and lightning were followed by the figures of ghosts, skeletons, and known individuals, whose eyes and mouth were made to move by the shifting of combined sliders. After the first figure had been exhibited for a short time, it began to grow less and less, as if removed to a great distance, and at last vanished in a small cloud of light. Out of this same cloud the germ of another figure began to appear, and gradually grew larger and larger, and approached the spectators till it attained its perfect developement. In this manner, the head of Dr Franklin was transformed into a skull; figures which retired with the freshness of life came back in the form of skeletons, and the retiring skeletons returned in the drapery of flesh and blood.

The exhibition of these transmutations was followed by spectres, skeletons, and terrific figures, which, instead of receding and vanishing as before, suddenly advanced upon the spectators, becoming larger as they approached them, and finally vanished by appearing to sink into the ground. The effect of this part of the exhibition was naturally the most impressive. The spectators were not only surprised but agitated, and many of them were of opinion that they could have touched the figures. M. Robertson at Paris introduced along with his pictures the direct shadows of living objects, which imitated

F

coarsely the appearance of those objects in a dark night or in moonlight.

All these phenomena were produced by varying the distance of the magic lantern A B, Fig. 5, from the screen P Q, which remained fixed, and at the same time keeping the image upon the screen distinct, by increasing the distance of the lens D from the sliders in E F. When the lantern approached to P Q, the circle of light P Q, or the section of the cone of rays P D Q, gradually diminished, and resembled a small bright cloud, when D was close to the screen. At this time a new figure was put in, so that when the lantern receded from the screen, the old figure seemed to have been transformed into the new one. Although the figure was always at the same distance from the spectators, yet, owing to its gradual diminution in size, it necessarily appeared to be retiring to a distance. When the magic lantern was withdrawn from P Q, and the lens D at the same time brought nearer to E F, the image in P Q gradually increased in size, and therefore seemed in the same proportion to be approaching the spectators.

Superior as this exhibition was to any representation that had been previously made by the magic lantern, it still laboured under several imperfections. The figures were poorly drawn, and in other respects not well executed, and no attempt whatever was made to remove the optical incongruity of the figures becoming more luminous when they retired from the observer, and more obscure when they approached to him. The variation of the distance of the lens D from the sliders in E F was not exactly adapted to the motion of the lantern to and from the screen, so that the outline of the figures was not equally distinct during their variations of magnitude.

Dr Thomas Young suggested the arrangement shown in Fig. 6 for exhibiting the phantasmagoria.

Fig. 6.

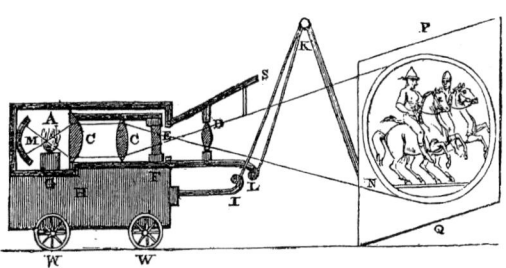

The magic lantern is mounted on a small car H, which runs on wheels W W.   The direct light of the lamp G, and that reflected from the mirror M, is condensed by the illuminating lenses C C, upon the transparent figures in the opaque sliders at E, and the image of these figures is formed at P Q, by the object lens D.   When the car H is drawn back on its wheels, the rod I K brings down the point K, and, by means of the rod K L, pushes the lens D nearer to the sliders in E F, and when the car advances to P Q, the point K is raised, and the rod K L draws out the lens D from the slider, so that the image is always in the conjugate focus of D, and therefore distinctly painted on the screen. The rod K N must be equal in length to I K, and

the point I must be twice the focal length of the
lens D before the object, L being immediately under the focus of the lens.   In order to diminish
the brightness of the image when it grows small
and appears remote, Dr Young contrived that the
support of the lens D should suffer a screen S to
fall and intercept a part of the light.   This method,
however, has many disadvantages, and we are satisfied, that the only way of producing a variation
in the light corresponding to the variation in the
size of the image, is to use a single illuminating lens
C, and to cause it to approach E F, and throw less
light upon the figures when D is removed from
E F, and to make C recede from E F when D approaches to it.   The lens C should therefore be
placed in a mean position corresponding to a
mean distance of the screen and to the ordinary
size of the figures, and should have the power of being removed from the slider E F, when a greater
intensity of light is required for the images when
they are rendered gigantic, and of being brought
close to E F when the images are made small.
The size of the lens C ought of course to be such
that the section of its cone of rays at E F is equal
to the size of the figure on the slider when C is at
its greatest distance from the slider.

The method recommended by Dr Young for pulling out and pushing in the object lens D, according
as the lantern approaches to or recedes from the
screen, is very ingenious and effective.   It is, however, clumsy in itself, and the connexion of the
levers with the screen, and their interposition between it and the lantern, must interfere with the
operations of the exhibitor.   It is, besides, suited

only to short distances between the screen and the lantern ; for when that distance is considerable, as it must sometimes require to be, the levers K L, K I, K T would bend by the least strain, and become unfitted for their purpose. For these reasons, the mechanism which adjusts the lens D should be moved by the axle of the front wheels, the tube which contains the lens should be kept at its greatest distance from E F by a slender spring, and should be pressed to its proper distance by the action of a spiral cam suited to the optical relation between the two conjugate focal distances of the lens.

Superior as the representations of the phantasmagoria are to those of the magic lantern, they are still liable to the defect which we have mentioned, namely, the necessary imperfection of the minute transparent figures when magnified. This defect cannot be remedied by employing the most skilful artists. Even Michael Angelo would have failed in executing a figure an inch long with transparent varnishes, when all its imperfections were to be magnified. In order, therefore, to perfect the art of representing phantasms, the objects must be living ones, and in place of chalky ill drawn figures mimicking humanity by the most absurd gesticulations, we shall have phantasms of the most perfect delineation, clothed in real drapery, and displaying all the movements of life. The apparatus by which such objects may be used may be called the *catadioptrical phantasmagoria,* as it operates both by reflexion and refraction.

The combination of mirrors and lenses which seems best adapted for this purpose is shown in Fig. 7, where A B, is a living figure placed before a large concave mirror M N, by means of which a diminished and inverted image of it is formed at *a b.* If P Q is the transparent screen upon which

Fig. 7.

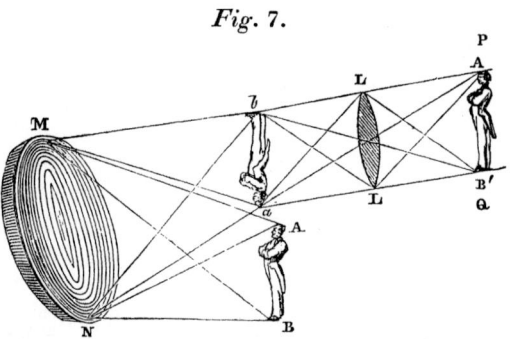

the image is to be shown to the spectators on the right hand of it, a large lens L L must be so placed before the image *a b* as to form a distinct and erect picture of it at A' B' upon the screen. When the image A' B' is required to be the exact size of A B, the lens L L must magnify the small image *a b* as much as the mirror M N diminishes the figure A B. The living object A B, the mirror M N, and the lens L L, must.all be placed in a moveable car for the purpose of producing the variations in the size of the phantasms, and the transformations of one figure into another. The contrivance for adjusting the lens L L to give a distinct picture at different distances of the screen will of course be required in the present apparatus. In order to give full effect to the phantasms, the living objects at A B will require to be illuminated in the strong-

est manner, and should always be dressed either
in white or in very luminous colours, and in order
to give them relief a black cloth should be stretched at some distance behind them.   Many interesting effects might also be produced by introducing
at A B fine paintings and busts.

It would lead us into too wide a field were we to
detail the immense variety of resources which the
science of optics furnishes for such exhibitions.
One of these, however, is too useful to be passed
without notice.   If we interpose a prism with a
small refracting angle between the image $a\ b$, Fig.
7, and the lens L L, the part of the figure immediately opposite to the prism will be as it were detached from the figure, and will be exhibited separately on the screen P Q.   Let us suppose that
this part is the head of the figure.   It may be detached vertically, or lifted from the body as if it
were cut off, or it may be detached downwards and
placed on the breast as if the figure were deformed.
In detaching the head vertically or laterally, an
opaque screen must be applied to prevent any part
of the head from being seen by rays which do not
pass through the prism; but this and other practical
details will soon occur to those who put the method
to an experimental trial.   The application of the
prism is shown in Fig. 8, where $a\ b$ is the inverted
image formed by a concave mirror, A B C a prism
with a small refracting angle B C A, placed between
$a\ b$ and the lens L L, s a small opaque screen, and
A B the figure with its head detached.   A hand
may be made to grasp the hair of the head, and the
aspect of death may be given to it, as if it had been
newly cut off.   Such a representation could be easily

made, and the effect upon the spectators would be quite overpowering. The lifeless head might then be made to recover its vitality, and be safely replaced upon the figure. If the head A of the living

Fig. 8.

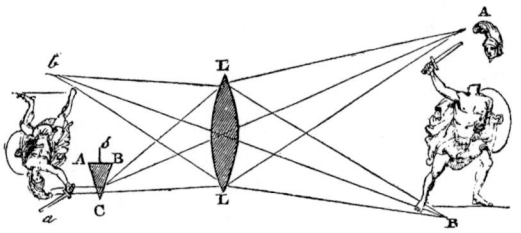

object A B, Fig. 7, is covered with black cloth, the head of a person or of an animal placed above A might be set upon the shoulders of the figure A B by the refraction of a prism.

When the figure *a b*, Fig. 8, is of very small dimensions, as in the magic lantern, a small prism of glass would answer the purpose required of it; but in public exhibitions where the image *a b* must be of a considerable size, if formed by a concave mirror, a very large prism would be necessary. This, however, though impracticable with solid glass, may be easily obtained by means of two large pieces of plate glass made into a prismatic vessel and filled with water. Two of the glasses of a carriage window would make a prism capable of doubling the whole of the bust of a living person placed as an object at A B, Fig. 7, so that two perfectly similar phantasms might be exhibited. In those cases

where the images before the lens L L are small,
they may be doubled and even tripled by interpo-
sing a well prepared plate of calcareous spar, that
is, crossed by a thin film. These images would
possess the singular character of being oppositely
coloured and of changing their distances and their
colours by slight variations in the positions of the
plate.*

In order to render the images which are formed
by the glass and water prisms as perfect as possible,
it would be easy to make them achromatic, and the
figures might be multiplied to any extent by using
several prisms, having their refracting edges parallel, for the purpose of giving a similarity of position
to all the images.

Among the instruments of natural magic which
were in use at the revival of science, there was one
invented by Kircher for exhibiting the mysterious

Fig. 9.

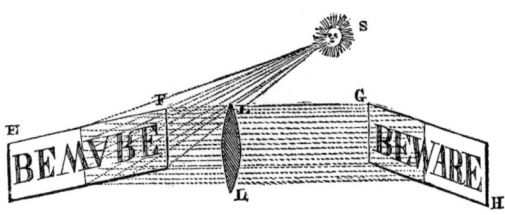

handwriting on the wall of an apartment from which
the magician and his apparatus were excluded. The
annexed figure represents this apparatus, as given

* See Edin. Encyclopædia.   Art. Optics, Vol. xv. p. 611.

by Schottus. The apartment in which the spectators are placed is between L L and G H, and there is an open window in the side next L L, G H being the inside of the wall opposite to the window. Upon the face of the plane speculum E F are written the words to be introduced, and when a lens L L is placed at such a distance from the speculum, and of such a focal length that the letters and the place of their representation are in its conjugate foci, a distinct image of the writing will be exhibited on the wall at G H. The letters on the speculum are of course inverted, as seen at E F, and when they are illuminated by the sun's rays S, as shown in the figure, a distinct image, as Schottus assures us, may be formed at the distance of 500 feet. In this experiment the speculum is by no means necessary. If the letters are cut out of an opaque card, and illuminated by the light of the sky in the day, or by a lamp during night, their delineation on the wall would be equally distinct. In the day time it would be necessary to place the letters at one end of a tube or oblong box, and the lens at the other end. As this deception is performed when the spectators are unprepared for any such exhibition, the warning written in luminous letters on the wall, or any word associated with the fate of the individual observer, could not fail to produce a singular effect upon his mind. The words might be magnified, diminished, multiplied, coloured and obliterated, in a cloud of light, from which they might again reappear by the methods already described, as applicable to the magic lantern.

The art of forming aërial representations was a great desideratum among the opticians of the 17th century. Vitellio and others had made many unsuccessful attempts to produce such images, and the speculations of Lord Bacon on the subject are too curious to be withheld from the reader.

" It would be well bolted out," says he, "whether great refractions may not be made upon reflexions, as well as upon direct beams. For example, take an empty basin, put an angel or what you will into it ; then go so far from the basin till you cannot see the angel, because it is not in a right line ; then fill the basin with water, and you shall see it out of its place, because of the refraction.   To proceed, therefore, put a looking-glass into a basin of water.   I suppose you shall not see the image in a right line or at equal angles, but wide.   I know not whether this experiment may not be extended, so as you might see the image and not the glass, which, for beauty and strangeness, were a fine proof, for then you should see the image like a spirit in the air. As, for example, if there be a cistern or pool of water, you shall place over against it the picture of the devil, or what you will, so as that you do not see the water.   Then put a looking-glass in the water ; now if you can see the devil's picture aside, not seeing the water, it would look like the devil indeed.   They have an old tale in Oxford, that Friar Bacon walked between two steeples, which was thought to be done by glasses, when he walked upon the ground. "

Kircher also devoted himself to the production of such images, and he has given in the annexed figure his method of producing them.   At the bottom of a polished cylindrical vessel A B he placed a figure C D, which we presume must have been highly illuminated from below, and to the spectators

who looked into the vessel in an oblique direction
there was exhibited an image placed vertically in
the air as if it were ascending at the mouth of the
vessel. Kircher assures us that he once exhibited
in this manner a representation of the Ascension of
our Saviour, and that the images were so perfect
that the spectators could not be persuaded till they

Fig. 10.

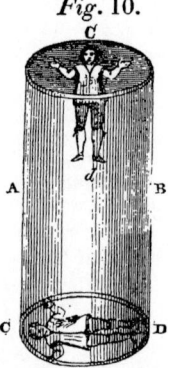

had attempted to handle them, that they were not
real substances. Although Kircher does not mention it, yet it is manifest that the original figure AB
must have been a deformed or anamorphous drawing,
in order to give a reflected image of just proportions. We doubt, indeed, if the representation or
the figure was ever exhibited. It is entirely incom-
patible with the laws of reflexion.

Among the ingenious and beautiful deceptions
of the 17th century, we must enumerate that of
the reformation of distorted pictures by reflexion
from cylindrical and conical mirrors. In these re-

presentations the original image from which a perfect picture is produced, is often so completely distorted, that the eye cannot trace in it the resemblance to any regular figure, and the greatest degree of wonder is of course excited, whether the
original image is concealed or exposed to view.
These distorted pictures may be drawn by strict
geometrical rules; but I have shown in Fig. 11 a

Fig. 11.

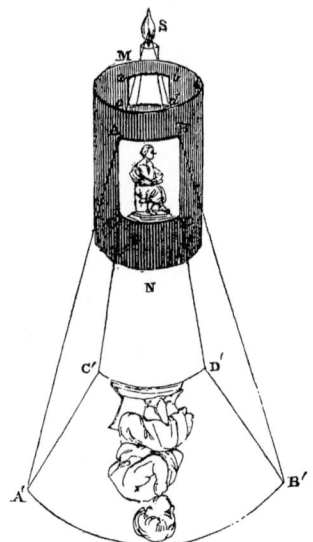

simple and practical method of executing them.
Let M N be an accurate cylinder made of tin-plate

or of thick pasteboard. Out of the farther side of
it cut a small aperture *a b c d* ; and out of the
nearer side cut a larger one A B C D, the size of
the picture to be distorted. Having perforated the
outline of the picture with small holes, place it on
the opening A B C D so that its surface may be cylindrical. Let a candle or a bright luminous object,
the smaller the better, be placed at S, as far behind
the picture A B C D as the eye is afterwards to
be placed before it, and the light passing through the
small holes will represent on a horizontal plane a
distorted image of the picture A′ B′ C′ D′, which
when sketched in outline with a pencil, and shaded
or coloured, will be ready for use. If we now sub-
stitute a polished cylindrical mirror of the same
size in place of M N, then the distorted picture

Fig. 12.

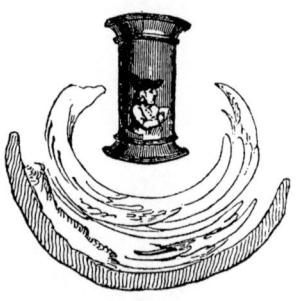

when laid horizontally at A′ B′ C′ D′ will be restored to its original state when seen by reflexion
at A B C D in the polished mirror. It would be

an improvement on this method to place at A B C D
a thin and flexible plate of transparent mica, having drawn upon it with a sharp point or painted
upon it the figure required.   The projected image
of this figure at A′ B′ C′ D′ may then be accurately copied.

The effect of a cylindrical mirror is shown in
Fig. 12, which is copied from an old one which we
have seen in use.

The method above described is equally applicable
to concave cylindrical mirrors, and to those of a
conical form, and it may also be applied to mirrors
of variable curvature, which produce different kinds
of distortions from different parts of their surfaces.

By employing a mirror whose surface has a variable curvature like A B C, Fig. 13, we obtain an

Fig. 13.

instrument for producing an endless variety of
caricatures, all of which are characterized by their

resemblance to the original. If a figure M N is
placed before such a mirror, it will of course appear distorted and caricatured, but even if the figure takes different distances and positions, the variations which the image undergoes are neither
sufficiently numerous or remarkable to afford much
amusement. But if the figure M N is very near
the mirror, so that new distortions are produced
by the different distances of its different parts from
the mirror, the most singular caricatures may be
exhibited. If the figure, for example, bends forward his head and the upper part of his body, they
will swell in size, leaving his lower extremities
short and slender. If it draws back the upper part
of the body and advances the limbs, the opposite
effect will take place. In like manner different
sides of the head, the right or the left side of it,
the brow or the chin, may be swelled and contracted at pleasure. By stretching out the arms
before the body they become like those of an
orang outang, and by drawing them back they
dwindle into half their regular size. All these
effects, which depend greatly on the agility and
skill of the performer, may be greatly increased by
suitable distortions in his own features and figure.
The family likeness, which is of course never lost
in all the variety of figures which are thus produced, adds greatly to the interest of the exhibition;
and we have seen individuals so annoyed at recognizing their own likeness in the hideous forms of
humanity which were thus delineated, that they
could not be brought to contemplate them a second
time. If the figure is inanimate, like the small
cast a of statue, the effect is very curious, as the

swelling and contracting of the parts and the sudden change of expression give a sort of appearance of vitality to the image. The inflexibility of such a figure, however, is unfavourable to its transformation into caricatures.

Interesting as these metamorphoses are, they lose in the simplicity of the experiment much of the wonder which they could not fail to excite if exhibited on a great scale, where the performer is invisible, and where it is practicable to give an aërial representation of the caricatured figures. This may be done by means of the apparatus shown in Fig. 7 * where we may suppose A B to be the reduced image seen in the reflecting surface A B C, Fig. 13.† By bringing this image nearer the mirror M N, Fig. 7, a magnified and inverted image of it may be formed at *a b*, of such a magnitude as to give the last image in P Q the same size as life. Owing to the loss of light by the two reflexions, a very powerful illumination would be requisite for the original figure. If such an exhibition were well got up the effect of it would be very striking.

Page 86.   † Page 95.

G

LETTER V.

Miscellaneous optical illusions—Conversions of cameos into intaglios, or elevations into depressions, and the reverse— Explanation of this class of deceptions—Singular effects of illumination with light of one simple colour—Lamps for producing homogeneous yellow light—Methods of increasing the effect of this exhibition—Method of reading the inscription of coins in the dark—Art of decyphering the effaced inscription of coins—Explanation of these singular effects— Apparent motion of the eyes in portraits—Remarkable examples of this—Apparent motion of the features of a portrait, when the eyes are made to move—Remarkable experiment of breathing light and darkness.

In the preceding letter I have given an account of the most important instruments of Natural Magic which depend on optical principles ; but there still remain several miscellaneous phenomena on which the stamp of the marvellous is deeply impressed, and the study of which is pregnant with instruction and amusement.

One of the most curious of these is that false perception in vision by which we conceive depressions to be elevations and elevations depressions, or by which intaglios are converted into cameos, and cameos into intaglios. This curious fact seems to have been first observed at one of the early meetings of the Royal Society of London, when one of the members,'in looking at a guinea through a compound microscope of new construction, was surprised to see the head upon the coin depressed, while other members could only see it embossed as it really was.

While using telescopes and compound microscopes, Dr Gmelin of Wurtemburg observed the same fact. The protuberant parts of objects appeared to him depressed, and the depressed parts protuberant; but what perplexed him extremely, this illusion took place at some times and not at others, in some experiments and not in others, and appeared to some eyes and not to others.

After making a great number of experiments, Dr Gmelin is said to have constantly observed the following effects: Whenever he viewed any object rising upon a plane of any colour whatever, provided it was neither white nor shining, and provided the eye and the optical tube were directly opposite to it, the elevated parts appeared depressed, and the depressed parts elevated. This happened when he was viewing a seal, and as often as he held the tube of the telescope perpendicularly, and applied it in such a manner that its whole surface almost covered the last glass of the tube. The same effect was produced when a compound microscope was used. When the object hung perpendicularly from a plane, and the tube was supported horizontally and directly opposite to it, the illusion also took place, and the appearance was not altered when the object hung obliquely and even horizontally. Dr Gmelin is said to have at last discovered a method of preventing this illusion, which was, by looking not towards the centre of the convexity, but at first to the edges of it only, and then gradually taking in the whole. " But why these things should so happen, he did not pretend to determine."

The best method of observing this deception, is to view the engraved seal of a watch with the eye-piece of an achromatic telescope, or with a compound microscope, or any combination of lenses, which inverts the objects that are viewed through it.* The depression in the seal will immediately appear an elevation, like the wax impression which is taken from it ; and though we know it to be hollow, and feel its concavity with the point of our finger, the illusion is so strong that it continues to appear a protuberance. The cause of this will be understood from Fig. 14, where S is the window of the apartment, or the light which illuminates

Fig. 14.

the *hollow* seal L R, whose shaded side is of course on the same side L with the light. If we now invert the seal with one or more lenses, so that it may look in the opposite direction, it will appear to the eye as in Fig. 15, with the shaded side L farthest from the window. But as we know that the

* A single convex lens will answer the purpose, provided we hold the eye six or eight inches behind the image of the seal formed in its conjugate focus.

window is still on our left hand, and that the light
falls in the direction R L, and as every body with
its shaded side farthest from the light must neces-

Fig. 15.

sarily be convex or protuberant, we immediately
believe that the hollow seal is now a cameo or
bas-relief. The proof which the eye thus receives
of the seal being raised, overcomes the evidence of
its being hollow derived from our actual knowledge,
and from the sense of touch. In this experiment
the deception takes place from our knowing the real
direction of the light which falls upon the seal :
for if the place of the window, with respect to the
seal, had been inverted as well as the seal itself,
the illusion could not have taken place.

In order to explain this better, let us suppose the
seal L R, Fig. 14, to be illuminated with a candle S,
the place of which we can change at pleasure. If
we invert L R it will rise into a cameo, as in Fig.
15 ; and if we then place another candle D on the
other side of it, as in Fig. 16, the hollow seal will
be equally illuminated on all sides, and it will sink
down into a cavity or intaglio. If the two candles
do not illuminate the seal equally, or if any accidental circumstance produces a belief that the light

is wholly or principally on one side, the mind will entertain a corresponding opinion respecting the state of the seal, regarding it as a hollow if it be-

Fig. 16.

lieves the light to come wholly or principally from the right hand, and as a cameo if it believes the light to come from the left hand.

If we use a small telescope to invert the seal, and if we cover up all the candle but its flame, and arrange the experiment so that the candle may be inverted along with the image, the seal will still retain its concavity, because the shadow is still on the same side with the illuminating body.

If we make the same experiments with the raised impression of the seal taken upon wax, we shall observe the very same phenomena, the seal being depressed when it alone is inverted, and retaining its convexity when the light is inverted along with it.

The illusion, therefore, under our consideration is the result of an operation of our own minds, whereby we judge of the forms of bodies by the knowledge we have acquired of light and shadow. Hence the illusion depends on the accuracy and extent of our knowledge on this subject; and while some persons are under its influence, others are entirely insensible to it. When the seal or hollow cavity is not polished but ground, and the surface round it of uniform colour and smoothness, almost every person, whether young or old, learned or ignorant, will be subject to the illusion; because the youngest and the most careless observers cannot but know that the shadow of a hollow is always on the side next the light, and the shadow of a protuberance on the side opposite to the light; but if the object is the raised impression of a seal upon wax, I have found that when inverted, it still seemed raised to the three youngest of six persons, while the three eldest were subject to the deception.

This illusion may be dissipated by a process of reasoning arising from the introduction of a new circumstance in the experiment. Thus, let R L Fig. 18, be the inverted seal, which consequently

Fig. 17.

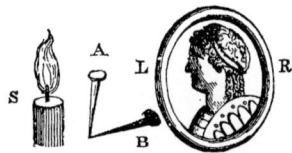

appears raised, and let an opaque and unpolished pin A be placed on one side of the seal. Its shadow will be of course opposite the candle as at B. In this case the seal, which had become a cameo by its inversion, will now sink down into a cavity by the introduction of the pin and its shadow; for as the pin and its shadow are inverted, as shown

in Fig. 18, while the candle retains its place, the
shadow of the pin falling in the direction A B is a
stronger proof to the eye that the light is coming
from the right hand, than the actual knowledge of

Fig. 18.

the candle being on the left hand, and therefore
the cameo necessarily sinks into a cavity, or the
shadow is now on the same side as the light. This
experiment will explain to us why on some occasions an acute observer will elude the deception,
while every other person is subject to it.   Let us
suppose that a particle of dust, or a little bit
of wax, capable of giving a shadow, is adhering
to the surface of the seal, an ordinary observer
will take no notice of this, or if he does, he will
probably not make it a subject of consideration,
and will therefore see the head on the seal raised
into a cameo; but the attentive observer noticing
the little protuberance, and observing that its shadow lies to the left of it, will instantly infer that
the light comes in that direction, and will still
see the seal hollow.

I have already mentioned that in some cases even
the sense of touch does not correct the erroneous
perception.   We of course feel that the part of
the hollow on which the finger is placed is actu-

ally hollow; but if we look at the other part of the hollow it will still appear raised.

By using two candles yielding different degrees of light, and thus giving an uncertainty to the direction of the light, we may weaken the illusion in any degree we choose, so as to overpower it by touch or by a process of reasoning.

I have had occasion to observe a series of analogous phenomena arising from the same cause, but produced without any instrument for inverting the object. If AB, for example, is a plate of mother of pearl, and LR a circular or any other cavity (Fig. 19,) ground or turned in it, then if this cavity is il-

Fig. 19.

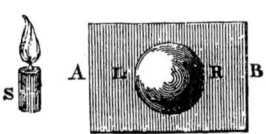

luminated by a candle or a window at S, in place of there being a shadow of the margin L of the hollow next the light, as there would have been had the body been opaque, a quantity of bright refracted light will appear where there would have been a shadow, and the rest of the cavity will be comparatively obscure, as if it were in shade. The necessary consequence of this is, that the cavity will appear as an elevation when seen only by the naked eye, as it is only an elevated surface that could have its most luminous side at L.

Similar illusions take place in certain pieces of

polished wood, calcedony, and mother of pearl, where
the surface is perfectly smooth.   This arises from
there being at that place a knot or growth, or no-
dule of different transparency from the surrounding
mass, and the cause of it will be understood from Fig.
20.   Let *m o* be the surface of a mahogany table,

Fig. 20.

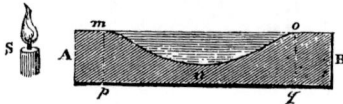

A *m o* B a section of the table, and *m n o*, a section
of a knot more transparent than the rest of the mass.
Owing to the transparency of the thin edge at *o*, opposite to the candle S, the side *o* is illuminated,
while the rest of the knot is comparatively dark,
so that on the principles already explained the spot
m n o appears to be a hollow in the table.   From
this cause arises the appearance of dimples in certain plates of calcedony called hammered calcedony, owing to its having the look of being dimpled
with a hammer   The surface on which these cavities are seen is a section of small spherical aggre-
gations of siliceous matter, which exhibit the same
phenomena as the cavities in wood.   Mother of
pearl presents the very same phenomena, and it is
indeed so common in this substance that it is nearly impossible to find a mother of pearl button or
counter which seems to have its surface flat, although
they are perfectly so when examined by the touch.

Owing to the different refraction of the incident light by the different growths of the shell cut in different directions by the artificial surface, like the annual growth of wood in a dressed plank, the surface has necessarily an unequal and undulating appearance.

Among the wonders of science there are perhaps none more surprising than the effects produced upon coloured objects by illuminating them with homogeneous light or light of one colour. The light which emanates from the sun, and by which all the objects of the material world are exhibited to us, is composed of three different colours, *red, yellow,* and *blue,* by the mixture of which in different proportions all the various hues of nature may be produced. These three colours, when mixed in the proportion in which they occur in the sun's rays, compose a purely white light; but if any body on which this white light falls shall absorb, or stop, or detain within its substance any part of any one or more of these simple colours, it will appear to the eye of that colour which arises from the mixture of all the rays which it does not absorb, or of that colour which white light would have if deprived of the colours which are absorbed. Scarlet cloth, for example, absorbs most of the blue rays and many of the yellow, and hence appears *red.* Yellow cloth absorbs most of the blue and many of the red rays, and therefore appears yellow, and blue cloth absorbs most of the yellow and red rays. If we were to illuminate the *scarlet* cloth with pure and unmixed *yellow* light, it would appear *yellow,* because the scarlet cloth does not absorb all the yellow rays, but reflects some of them ; and if we illuminate *blue* cloth with yellow

light, it will appear nearly *black*, because it absorbs
all the yellow light, and reflects almost none of it.
But whatever be the nature and colour of the bodies
on which the yellow light falls, the light which it
reflects must be yellow, for no other light falls upon
them, and those which are not capable of reflecting
yellow light must appear absolutely black, however
brilliant be their colour in the light of day.

As the methods now discovered of producing
yellow light in abundance were not known to the
ancient conjurors, nor even to those of later times,
they have never availed themselves of this valuable
resource.   It has been long known that salt thrown

Fig. 21.

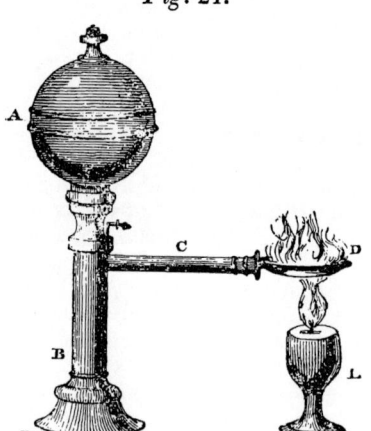

into the wick of a flame produces yellow light, but
this light is mixed with blue and green rays, and is,

besides, so small in quantity, that it illuminates objects only that are in the immediate vicinity of the
flame.   A method which I have found capable of
producing it in abundance is shown in Fig. 21,
where A B is a lamp containing at A a large quantity of alcohol and water or ardent spirits, which
gradually descends into a platina or metallic cup D.
This cup is strongly heated by a spirit-lamp L, inclosed in a dark lantern, and when the diluted alcohol in D is inflamed, it will burn with a fierce
and powerful yellow flame : If the flame should not
be perfectly yellow, owing to an excess of alcohol,
a proportion of salt thrown into the cup will answer the same purpose as a farther dilution of the
alcohol.*

A monochromatic lamp for producing yellow
light may be constructed most effectually by employing a portable gas lamp containing compressed
oil gas. If we allow the gas to escape in a copious
stream, and set it on fire, it will form an explosive
mixture with the atmospheric air, and will no longer burn with a white flame, but will emit a bluish and reddish light. The force of the issuing gas,
or any accidental current of air, is capable of blowing out this flame, so that it is necessary to have
a contrivance for sustaining it. The method which
I used for this purpose is shown in Fig. 22.   A
small gas tube $a\ b\ c$, arising from the main burner
M N of the gas lamp P Q, terminates above the
burner, and has a short tube $d\ e$, moveable up and
down within it, so as to be gas-tight.   This tube
$d\ e$, closed at e, communicates with the hollow ring
fg, in the inside of which four apertures are per-

* See *Edinburgh Transactions*, vol. ix. p. 435.

forated in such a manner as to throw their jets of
gas to the apex of a cone, of which *f g* is the base.
When we cause the gas to flow from the burner M,

Fig. 22.

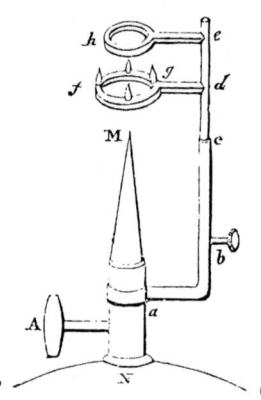

by opening the main cock A, it will rush into the
tube *a b c d*, and issue in small flames at the four
holes in the ring *f g*.   The size of these flames is
regulated by the cock *b*.   The inflammation, therefore, of the ignited gas will be sustained by these
four subsidiary flames through which it passes, independent of any agitation of the air, or of the
force with which it issues from the burner.   On a
projecting arm *e h*, carrying a ring *h*, I fixed a
broad collar, made of coarse cotton wick, which had
been previously soaked in a saturated solution of
common salt.   When the gas was allowed to escape
at M, with such force as to produce a long and
broad column of an explosive mixture of gas and

3

atmospheric air, the bluish flame occasioned by the explosion passes through the salted collar, and is converted by it into a mass of homogeneous yellow light. This collar will last a long time without any fresh supply of salt, so that the gas lamp will yield a permanent monochromatic yellow flame, which will last as long as there is gas in the reservoir. In place of a collar of cotton wick, a hollow cylinder of sponge, with numerous projecting tufts, may be used, or a collar may be similarly constructed with asbestos cloth, and, if thought necessary, it might be supplied with a saline solution from a capillary fountain.

Having thus obtained the means of illuminating any apartment with yellow light, let the exhibition be made in a room with furniture of various bright colours, with oil or water coloured paintings on the wall. The party which is to witness the experiment should be dressed in a diversity of the gayest colours; and the brightest coloured flowers and highly coloured drawings should be placed on the tables. The room being at first lighted with ordinary lights, the bright and gay colours of every thing that it contains will be finely displayed. If the white lights are now suddenly extinguished, and the yellow lamps lighted, the most appalling metamorphosis will be exhibited. The astonished individuals will no longer be able to recognize each other. All the furniture in the room and all the objects which it contains will exhibit only one colour. The flowers will lose their hues. The paintings and drawings will appear as if they were executed in China ink, and the gayest dresses, the brightest scarlets, the purest lilacs, the richest blues, and the most

vivid greens will all be converted into one monotonous yellow. The complexions of the parties too
will suffer a corresponding change. One pallid
death-like yellow,

> ———like the unnatural hue
> Which autumn plants upon the perished leaf,

will envelope the young and the old, and the sallow
faces will alone escape from the metamorphosis.
Each individual derives merriment from the cada-
verous appearance of his neighbour, without being
sensible that he is himself one of the ghostly assemblage.

If, in the midst of the astonishment which is
thus created, the white lights are restored at one end of
the room, while the yellow lights are taken to the
other end, one side of the dress of every person,
namely, that next the white light, will be restored
to its original colours, while the other side will retain its yellow hue. One cheek will appear in a
state of health and colour, while the other retains
the paleness of death, and, as the individuals change
their position, they will exhibit the most extraordinary transformations of colour.

If, when all the lights are yellow, beams of white
light are transmitted through a number of holes
like those in a sieve, each luminous spot will restore the colour of the dress or furniture upon
which it falls, and the nankeen family will appear
all mottled over with every variety of tint. If
a magic lantern is employed to throw upon the
walls or upon the dresses of the company luminous figures of flowers or animals, the dresses will
be painted with these figures in the real colour of

4

the dress itself. Those alone who appeared in yellow, and with yellow complexions, will to a great degree escape all these singular changes.

If red and blue light could be produced with the same facility and in the same abundance as yellow light, the illumination of the apartment with these lights in succession would add to the variety and wonder of the exhibition. The red light might perhaps be procured in sufficient quantity from the nitrate and other salts of strontian; but it would be difficult to obtain a blue flame of sufficient intensity for the suitable illumination of a large room. Brilliant white lights, however, might be used, having for screens glass troughs containing a mass one or two inches thick of a solution of the ammoniacal carbonate of copper. This solution absorbs all the rays of the spectrum but the blue, and the intensity of the blue light thus produced would increase in the same proportion as the white light employed.

Among the numerous experiments with which science astonishes and sometimes even strikes terror into the ignorant, there is none more calculated to produce this effect than that of displaying to the eye in absolute darkness the legend or inscription upon a coin. To do this, take a silver coin, (I have always used an old one,) and after polishing the surface as much as possible, make the parts of it which are raised rough by the action of an acid, the parts not raised, or those which are to be rendered darkest, retaining their polish. If the coin thus prepared is placed upon a mass of red hot iron, and removed into a dark room, the inscription upon it will become less luminous than the rest, so that it may be distinctly read by the spectator.

H

The mass of red hot iron should be concealed from the observer's eye, both for the purpose of rendering the eye fitter for observing the effect, and of removing all doubt that the inscription is really read in the dark, that is, without receiving any light, direct or reflected, from any other body. If, in place of polishing the depressed parts, and roughening its raised parts, we make the raised parts polished, and roughen the depressed parts, the inscription will now be less luminous than the depressed parts, and we shall still be able to read it, from its being as it were written in black letters on a white ground. The first time I made this experiment, without being aware of what would be the result, I used a French shilling of Louis XV and I was not a little surprised to observe upon its surface in black letters the inscription BENEDICTUM SIT NOMEN DEI.

The most surprising form of this experiment is when we use a coin from which the inscription has been either wholly obliterated, or obliterated in such a degree as to be illegible. When such a coin is laid upon the red hot iron, the letters and figures become oxidated, and the film of oxide radiating more powerfully than the rest of the coin will be more luminous than the rest of the coin, and the illegible inscription may be now distinctly read to the great surprise of the observer, who had examined the blank surface of the coin previous to its being placed upon the hot iron. The different appearances of the same coin, according as the raised parts are polished or roughened, are shown in Fig. 23 and 24.

In order to explain the cause of these remarkable effects, we must notice a method which has been long known, though never explained, of decyphering the inscriptions on worn out coins. This is done by merely placing the coin upon a hot iron :

Fig. 23. *Fig.* 24.

An oxidation takes place over the whole surface of the coin, the film of oxide changing its tint with the intensity or continuance of the heat. The parts, however, where the letters of the inscription had existed, oxidate at a different rate from the surrounding parts, so that these letters exhibit their shape, and become legible in consequence of the film of oxide which covers them having a different thickness, and therefore reflecting a different tint from that of the adjacent parts. The tints thus developed sometimes pass through many orders of brilliant colours, particularly *pink* and *green*, and settle in a bronze, and sometimes a black tint, resting upon the inscription alone. In some cases the tint left on the trace of the letters is so very faint that it can just be seen, and may be entirely removed by a slight rub of the finger.

When the experiment is often repeated with the same coin, and the oxidations successively removed after each experiment, the film of oxide continues to diminish, and at last ceases to make its appearance. It recovers the property, however, in the course of time. When the coin is put upon the hot iron, and consequently when the oxidation is the greatest, a considerable smoke arises from the coin, and this diminishes like the film of oxide by frequent repetition. A coin which had ceased to emit this smoke, smoked slightly after having been exposed twelve hours to the air. I have found from numerous trials that it is always the raised parts of the coin, and in modern coins the elevated ledge round the inscription, that become first oxidated. In an English shilling of 1816 this ledge exhibited a brilliant yellow tint before it appeared on any other part of the coin.

If we use an uniform and homogeneous disc of silver that has never been hammered or compressed, its surface will oxidate equally, provided all its parts are equally heated. In the process of converting this disc into a coin, the *sunk* parts have obviously been *most compressed* by the prominent parts of the die, and the *elevated* parts *least compressed*, the metal being in the latter left as it were in its natural state. The raised letters and figures on a coin have therefore less density than the other parts, and these parts oxidate sooner or at a lower temperature. When the letters of the legend are worn off by friction, the parts immediately below them have also less density than the surrounding metal, and the site as it were of the letters therefore receive from heat a degree of oxidation, and a colour different from that of the surrounding surface. Hence we obtain an explanation of the revival of the invisible letters by oxidation.

The same influence of difference of density may be observed in the beautiful oxidations which are produced on the surface of highly polished steel, heated in contact with air, at temperatures between 430° and 630° of Fahrenheit. * When the steel has hard portions called *pins* by the workmen, the uniform tint of the film of oxide stops near these hard portions, which always exhibit colours different from those of the rest of the mass. These parts, on account of their increased density, absorb the oxygen of atmospheric air less copiously than the surrounding portions. Hence we see the cause why steel expanded by heat absorbs oxygen, which, when united with the metal, forms the coloured superficial film. As the heat increases, a greater quantity of oxygen is absorbed, and the film increases in thickness.

These observations enable us to explain the legibility of inscriptions in the dark, whether the coin is in a perfect state, or the letters of it worn off. All *black* or *rough* surfaces radiate light more copiously than *polished* or *smooth* surfaces, and hence the inscription is *luminous* when it is *rough*, and *obscure* when it is polished, and the letters covered with black oxide are more luminous than the adjacent parts, on account of the superior

* See *Edinburgh Encyclopædia*, Art. Steel, vol. xviii. p. 387.

radiation of light by the black oxide which covers them.

By the means now described invisible writing might be conveyed by impressing it upon a metallic surface, and afterwards erasing it by grinding and polishing that surface perfectly smooth. When exposed to a proper degree of heat, the secret would display itself written in oxidated letters. Many amusing experiments might be made upon the same principle.

A series of curious and sometimes alarming deceptions, arises from the representation of objects in perspective upon a plane surface. One of the most interesting of these depends on the principles which regulate the apparent direction of the eyes in a portrait. Dr Wollaston has thought this subject of sufficient importance to be treated at some length in the Philosophical Transactions. When we look at any person we direct to them both our face and our eyes, and in this position the circular iris will be in the middle of the white of the eyeball, or, what is the same thing, there will be the same quantity of white on each side of the iris. If the eyes are now moved to either side, while the head remains fixed, we shall readily judge of the change of their direction by the greater or less quantity of white on each side of the iris. This test, however, accurate as it is, enables us only to estimate the extent to which the eyes deviate in direction from the direction of the face to which they belong. But their direction in reference to the person who views them is entirely a different matter ; and Dr Wollaston is of opinion, that we are not guided by the eyes alone, but are unconsciously aided by the concurrent position of the entire face.

If a skilful painter draws a pair of eyes with great correctness directed to the spectator, and deviating from the general position of the face as much as is usual in good portraits, it is very difficult to determine their direction, and they will appear to have different directions to different persons. But what is very curious, Dr Wollaston has shown that the same pair of eyes may be made to direct themselves either to or from the spectator by the addition of other features in which the position of the face is changed. Thus in Fig. 25, the pair of

Fig. 25.

Fig. 26.

eyes are looking intently at the spectator, and the face has a corresponding direction; but when we cover up the face in Fig. 25 with the face in Fig. 26, which looks to the right, the eyes change their direction, and look to the right also. In like man-

ner, eyes drawn originally to look a little to the
right or the left of the spectator may be made to
look directly at him by adding suitable features.

The nose is obviously the principal feature which
produces this change of direction, as it is more subject to change of perspective than any of the other
features ; but Dr Wollaston has shown by a very
accurate experiment, that even a small portion of
the nose introduced with the features will carry
the eyes along with it. He obtained four exact
copies of the same pair of eyes looking at the spectator, by transferring them upon copper from a
steel plate, and having added to each of two pair of
them, a nose in one case directed to the right,
and in the other, to the left, and to each of the
other two pairs a very small portion of the upper part
of the nose, all the four pair of eyes lost their front
direction, and looked to the right or to the left,
according to the direction of the nose, or of the
portion of it which was added.

But the effect thus produced is not limited, as
Dr Wollaston remarks, to the mere change in the
direction of the eyes, " for a total difference of character may be given to the same eyes by a due representation of the other features. A lost look of
devout abstraction in an uplifted countenance may
be exchanged for an appearance of inquisitive archness in the leer of a younger face turned downwards
and obliquely towards the opposite side," as in Fig.
27, 28. This, however, is perhaps not an exact expression of the fact. The new character which is said
to be given to the eyes is given only to the eyes in
combination with the new features, or, what is pro-

bably more correct, the inquisitive archness is in the other features, and the eye does not belie it.

Dr Wollaston has not noticed the converse of these illusions, in which a change of direction is

Fig. 27.

Fig. 28.

given to fixed features by a change in the direction of the eyes. This effect is finely seen in some magic lantern sliders, where a pair of eyes is made to move in the head of a figure which invariably follows the motion of the eyeballs.

Having thus determined the influence which the general perspective of the face has upon the apparent direction of the eyes in a portrait, Dr Wollaston applies it to the explanation of the well known fact, that when the eyes of a portrait look at a spectator in front of it they will follow him, and appear to look at him in every other direction.   This curious fact, which has received less consideration than it merits, has been often skilfully employed by the novelist, in alarming the fears or exciting the courage of his hero.   On returning to the hall of his ancestors, his attention is powerfully fixed on the grim portraits which surround him.   The parts which they have respectively performed in the family history rise to his mind : his own actions, whether good or evil, are called up in contrast, and as the preserver or the destroyer of his line, he stands as it were in judgment before them.   His imagination, thus excited by conflicting feelings, transfers a sort of vitality to the canvas, and if the personages do not " start from their frames," they will at least bend upon him their frowns or their approbation.   It is in vain that he tries to evade their scrutiny.   Wherever he goes their eyes eagerly pursue him ;—they will seem even to look at him over their shoulders, and he will find it impossible to shun their gaze but by quitting the apartment.

As the spectator in this case changes his position in a horizontal plane, the effect which we have described is accompanied by an apparent diminution in the breadth of the human face, from only seven or eight inches till it disappears at a great obliquity.   In moving, therefore, from a front view to the most oblique view of the face, the change in its apparent breadth is so slow that the apparent motion of the head of the figure is scarcely recognized as it follows the spectator.   But if the perspective

figure has a great breadth in a horizontal plane,
such as a soldier firing his musket, an artilleryman
his piece of ordnance, a bowman drawing his bow,
or a lancer pushing his spear, the apparent breadth
of the figure will vary from five to six feet or upwards till it disappears, and therefore the change of
apparent magnitude is sufficiently rapid to give the
figure the dreaded appearance of turning round, and
following the spectator. One of the best examples
of this must have been often observed in the fore-
shortened figure of a dead body lying horizontally,
which has the appearance of following the observer
with great rapidity, and turning round upon the
head as the centre of motion.

The cause of this phenomenon is easily explained. Let us suppose a portrait with its face and its
eyes directed straight in front, so as to look at the
spectator. Let a straight line be drawn through the
tip of the nose and half way between the eyes, which
we shall call the middle line. On each side of this
middle line there will be the same breadth of head,
of cheek, of chin, and of neck, and each iris will
be in the middle of the whole of the eye. If we
now go to one side, the apparent horizontal breadth
of every part of the head and face will be diminished, but the parts on each side of the middle line
will be diminished equally, and at any position,
however oblique, there will be the same breadth of
face on each side of the middle line, and the iris
will be in the centre of the whole of the eyeball, so
that the portrait preserves all the characters of a
figure looking at the spectator, and must necessarily do so wherever he stands.

This explanation might be illustrated by a picture

which represents three artillerymen, each firing a piece of ordnance in parallel directions. Let the gun of the middle one be pointed accurately to the eye of the spectator, so that he sees neither its right side nor its left, nor its upper nor its under side, but directly down its muzzle, so that if there was an opening in the breech he would see through it. In like manner the spectator will see the left side of the gun on his left hand, and the right side of the gun on his right hand. If the spectator now changes his place, and takes ever such an oblique position, either laterally or vertically, he must still see the same thing, because nothing else is presented to his view. The gun of the middle soldier must always point to his eye, and the other guns to the right and left of him. They must therefore all three seem to move as he moves, and follow his eye in all its changes of place. The same observations are of course applicable to buildings and streets seen in perspective.

In common portraits the apparent motion of the head is generally rendered indistinct by the canvas being imperfectly stretched, as the slightest concavity or convexity entirely deforms the face when the obliquity is considerable. The deception is therefore best seen when the painting is executed on a very flat board, and in colours sufficiently vivid to represent every line in the face with tolerable distinctness at great obliquities. This distinctness of outline is indeed necessary to a satisfactory exhibition of this optical illusion. The most perfect exhibition, indeed, that I ever saw of it was in the case of a painting of a ship upon a sign-board executed in strongly gilt lines. It contained a view of the

stern, and side of a ship in the stocks, and, owing to
the flatness of the board and the brightness of the
lines, the gradual developement of the figure from the
most violent foreshortening at great obliquities till
it attained its perfect form, was an effect which surprised every person that saw it.

The only other optical illusion which our limits
will permit us to explain, is the very remarkable
experiment of what may be truly called *breathing
light or darkness*. Let S be a candle where light
falls at an angle of 56° 45′ upon two glass plates
A, B, placed close to each other, and let the reflected rays A C, B D, fall at the same angle upon two
similar plates C, D, but so placed that the plane of
reflexion from the latter is at right angles to the
plane of reflexion from the former. An eye placed

Fig. 29.

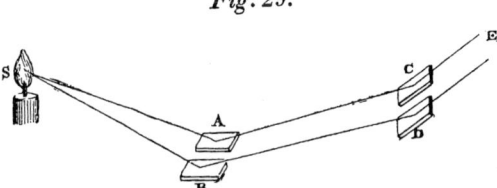

at E, and looking at the same time into the two plates
C and D, will see very faint images of the candle
S, which by a slight adjustment of the plates may
be made to disappear almost wholly. Allowing the
plate C to remain as it is, change the position of
D, till its inclination to the ray BD is diminished
about $3\frac{1}{2}$°, or made nearly 53° 11′. When this is
done, the image that had disappeared on looking into
D will be restored, so that the spectator at E, upon

looking into the two mirrors C, D, will see no light in C, because the candle has nearly disappeared, while the candle is distinctly seen in D. If, while the spectator is looking into these two mirrors, either he or another person breathes upon them gently and quickly, the breath will revive the extinguished image in C, and will extinguish the visible image in D. The following is the cause of this singular result. The light AC, BD, is polarized by reflexion from the plates A, B, because it is incident at the polarizing angle of 56° 45' for glass. When we breathe upon the plates C, D, we form upon their surface a thin film of water, whose polarizing angle is 53° 11 , so that if the polarized rays AC, BD, fell upon the plates C, D, at an angle of 53° 11', the candle from which they proceeded would not be visible, or they would not suffer reflexion from the plates C D. At all other angles the light would be reflected and the candles visible. Now the plate D is placed at an angle of 53° 11' and C at an angle of 56° 45', so that when a film of water is breathed upon them the light will be reflected from the latter, and none from the former: that is, the act of breathing upon the plates will restore the invisible, and extinguish the visible image.

LETTER VI.

AMONG the wonders of the natural world which are every day presented to us, without either exciting our surprise or attracting our notice, some are occasionally displayed which possess all the characters of supernatural phenomena. In the names by which they are familiarly known, we recognize the terror which they inspired, and even now, when science has reduced them to the level of natural phenomena, and developed the causes from which they arise, they still retain their primitive importance, and are watched by the philosopher with as intense an interest as when they were deemed the immediate effects of Divine power. Among these phenomena we may enumerate the *Spectre*

of the Brocken, the *Fata Morgana* of the Straits
of Messina, the *Spectre Ships* which appear in the
air, and the other extraordinary effects of the *Mirage*.*

The Brocken is the name of the loftiest of the
Hartz Mountains, a picturesque range which lies
in the kingdom of Hanover. It is elevated 3300
feet above the sea, and commands the view of a
plain seventy leagues in extent, occupying nearly
the 200dth part of the whole of Europe, and animated with a population of above five millions of
inhabitants. From the earliest periods of authen-
tic history, the Brocken has been the seat of the
marvellous. On its summits are still seen huge
blocks of granite, called the Sorcerer's Chair and
the Altar. A spring of pure water is known by the
name of the Magic Fountain, and the Anemone of
the Brocken is distinguished by the title of the
Sorcerer's Flower. These names are supposed to
have originated in the rites of the great Idol Cor-
tho, whom the Saxons worshipped in secret on the
summit of the Brocken, when Christianity was
extending her benignant sway over the subjacent
plains.

As the locality of these idolatrous rites, the
Brocken must have been much frequented, and we
can scarcely doubt that the spectre which now so
often haunts it at sunrise must have been observ-

In the Sanscrit, says Baron Humboldt, the phenomenon of the Mirage is called *Mriga Trichna* " thirst or desire
of the antelope," no doubt because this animal *Mriga*, compelled by thirst, *Trichna*, approaches those barren plains
where, from the effect of unequal refraction, he thinks he
perceives the undulating surface of the waters.—*Personal
Narrative*, Vol. iii. p. 554.

ed from the earliest times ; but it is nowhere mentioned that this phenomenon was in any way associated with the objects of their idolatrous worship.
One of the best accounts of the spectre of the
Brocken is that which is given by M. Haue, who
saw it on the 23d of May 1797. After having
been on the summit of the mountain no less than
thirty times, he had at last the good fortune of
witnessing the object of his curiosity. The sun rose
about four o'clock in the morning through a serene
atmosphere. In the south-west, towards Achter-
mannshohe, a brisk west wind carried before it the
transparent vapours, which had not yet been condensed into thick heavy clouds. About a quarter
past four he went towards the inn, and looked
round to see whether the atmosphere would afford
him a free prospect towards the south-west, when
he observed at a very great distance, towards Ach-
termannshohe, a human figure of a monstrous size.
His hat, having been almost carried away by a violent gust of wind, he suddenly raised his hand to
his head, to protect his hat, and the colossal figure
did the same. He immediately made another
movement by bending his body,—an action which
was repeated by the spectral figure. M. Haue was
desirous of making farther experiments, but the figure disappeared. He remained, however, in the
same position, expecting its return, and in a few
minutes it again made its appearance on the Ach-
termannshohe, when it mimicked his gestures as
before. He then called the landlord of the inn,
and having both taken the same position which he
had before, they looked towards the Achtermanns-
hohe but saw nothing. In a very short space of

I

time, however, two colossal figures were formed
over the above eminence, and after bending their
bodies and imitating the gestures of the two spectators, they disappeared. Retaining their position,
and keeping their eyes still fixed upon the same
spot, the two gigantic spectres again stood before
them, *and were joined by a third.* Every movement that they made was imitated by the three figures, but the effect varied in its intensity, being
sometimes weak and faint, and at other times strong
and well defined.

Fig. 30.

In the year 1798 M. Jordan saw the same phenomenon at sunrise, and under similar circumstances, but with less distinctness, and without any duplication of the figures.*

Phenomena perfectly analogous to the preceding,
though seen under less imposing circumstances,

* See J. F. Gmelin's *Gottingischen Journal der Wissen-
chaften,* Vol. i. part iii. 1798.

have been often witnessed. When the spectator
sees his own shadow opposite to the sun upon a mass
of thin fleecy vapour passing near him, it not only
imitates all his movements, but its head is distinctly encircled with a halo of light. The aërial figure
is often not larger than life, its size and its apparent distance depending, as we shall afterwards see,
upon particular causes. I have often seen a similar shadow when bathing in a bright summer's day
in an extensive pool of deep water. When the
fine mud deposited at the bottom of the pool is
disturbed by the feet of the bather, so as to be dis-
seminated through the mass of water in the direction of his shadow, his shadow is no longer a shapeless mass formed upon the bottom, but is a regular
figure formed upon the floating particles of mud,
and having the head surrounded with a halo, not
only luminous, but consisting of distinct radiations.

One of the most interesting accounts of aërial
spectres with which we are acquainted has been
given by Mr James Clarke, in his Survey of the
Lakes of Cumberland, and the accuracy of this account was confirmed by the attestations of two of
the persons by whom the phenomena were first
seen. On a summer's evening in the year 1743,
when Daniel Stricket, servant to John Wren of
Wilton Hall, was sitting at the door along with
his master, they saw the figure of a man with a
dog pursuing some horses along Souterfell side, a
place so extremely steep that a horse could scarcely travel upon it at all. The figures appeared to
run at an amazing pace, till they got out of sight
at the lower end of the fell. On the following
morning Stricket and his master ascended the steep

side of the mountain, in the full expectation of
finding the man dead, and of picking up some of the
shoes of the horses, which they thought must have
been cast while galloping at such a furious rate.
Their expectations, however, were disappointed.
No traces either of man or horse could be found,
and they could not even discover upon the turf the
single mark of a horse's hoof. These strange appearances, seen at the same time by two different
persons in perfect health, could not fail to make a
deep impression on their minds. They at first concealed what they had seen, but they at length disclosed it, and were laughed at for their credulity.

In the following year, on the 23d June 1744,
Daniel Stricket, who was then servant to Mr Lancaster of Blakehills, (a place near Wilton Hall, and
both of which places are only about half a mile
from Souterfell,) was walking, about seven o'clock
in the evening, a little above the house, when he
saw a troop of horsemen riding on Souterfell side
in pretty close ranks, and at a brisk pace. Recollecting the ridicule that had been cast upon him
the preceding year, he continued to observe the figures for some time in silence; but being at last
convinced that there could be no deception in the
matter, he went to the house and informed his
master that he had something curious to show him.
They accordingly went out together; but before
Stricket had pointed out the place Mr Lancaster's son had discovered the aërial figures. The
family was then summoned to the spot, and the
phenomena were seen alike by them all. The eques-
trian figures seemed to come from the lowest parts
of Souterfell, and became visible at a place called

Knott. They then advanced in regular troops along the side of the Fell, till they came opposite to Blakehills, when they went over the mountain, after describing a kind of curvilineal path. The pace at which the figures moved was a regular swift walk, and they continued to be seen for upwards of two hours, the approach of darkness alone preventing them from being visible. Many troops were seen in succession; and frequently the last but one in a troop quitted his position, galloped to the front, and took up the same pace with the rest. The changes in the figures were seen equally by all the spectators, and the view of them was not confined to the farm of Blakehills only, but they were seen by every person at every cottage within the distance of a mile, the number of persons who saw them amounting to about twenty-six. The attestation of these facts, signed by Lancaster and Stricket, bears the date of the 21st July 1785.

These extraordinary sights were received not only with distrust but with absolute incredulity. They were not even honoured with a place in the records of natural phenomena, and the philosophers of the day were neither in possession of analogous facts, nor were they acquainted with those principles of atmospherical refraction upon which they depend. The strange phenomena, indeed, of the *Fata Morgana,* or the *Castles of the Fairy Morgana,* had been long before observed, and had been described by Kircher in the 17th century, but they presented nothing so mysterious as the aërial troopers of Souterfell; and the general characters of the two phenomena were so unlike, that even a philosopher might have been excused for ascribing them to different causes.

This singular exhibition has been frequently seen in the Straits of Messina between Sicily and the coast of Italy, and whenever it takes place, the people, in a state of exultation, as if it were not only a pleasing but a lucky phenomenon, hurry down to the sea, exclaiming *Morgana, Morgana.* When the rays of the rising sun form an angle of 45° on the sea of Reggio, and when the surface of the water is perfectly unruffled either by the wind or the current, a spectator placed upon an eminence in the city, and having his back to the sun and his face to the sea, observes upon the surface of the water superb palaces with their balconies and windows, lofty towers, herds and flocks grazing in wooded vallies and fertile plains, armies of men on horseback and on foot, with multiplied fragments of buildings, such as columns, pilasters, and arches. These objects pass rapidly in succession along the surface of the sea during the brief period of their appearance. The various objects thus enumerated are pictures of palaces and buildings actually existing on shore, and the living objects are of course only seen when they happen to form a part of the general landscape.

If at the time that these phenomena are visible the atmosphere is charged with vapour or dense exhalations, the same objects which are depicted upon the sea will be seen also in the air occupying a space which extends from the surface to the height of twenty-five feet. These images, however, are less distinctly delineated than the former.

If the air is in such a state as to deposit dew, and is capable of forming the rainbow, the objects will be seen only on the surface of the sea, but they all appear fringed with red, yellow, and blue light as if they were seen through a prism.

In our own country, and in our own times, facts still more extraordinary have been witnessed. From Hastings, on the coast of Sussex, the cliffs on the French coast are fifty miles distant, and they are actually hid by the convexity of the earth, that is, a strait line drawn from Hastings to the French coast would pass through the sea. On Wednesday the 26th July 1798, about five o'clock in the afternoon, Mr Latham, a Fellow of the Royal Society, then residing at Hastings, was surprised to see a crowd of people running to the sea side. Upon inquiry into the cause of this, he learned that the coast of France could be seen by the naked eye, and he immediately went down to witness so singular a sight. He distinctly saw the cliffs extending for some leagues along the French coast, and they appeared as if they were only a few miles off. They gradually appeared more and more elevated, and seemed to approach nearer to the eye. The sailors with whom Mr Latham walked along the water's edge were at first unwilling to believe in the reality of the appearance, but they soon became so thoroughly convinced of it, that they pointed out and named to him the different places which they had been accustomed to visit, and which they conceived to be as near as if they were sailing at a small distance into the harbour. These appearances continued for nearly an hour, the cliffs sometimes appearing brighter and nearer, and at other times fainter and more remote. Mr Latham then went upon the eastern cliff or hill, which is of considerable

height, when, as he remarks, a most beautiful
scene presented itself to his view. He beheld at
once Dungeness, Dover Cliffs, and the French coast
all along from Calais, Boulogne, &c. to St Vallery,
and, as some of the fishermen affirmed, as far west
as Dieppe. With the help of a telescope, the French
fishing boats were plainly seen at anchor, and the
different colours of the land upon the heights, together with the buildings, were perfectly discernible.
Mr Latham likewise states that the cape of land
called Dungeness, which extends nearly two miles
into the sea, and is about sixteen miles in a straight
line from Hastings, appeared as if quite close to it,
and the vessels and fishing-boats which were sailing between the two places appeared equally near,
and were magnified to a high degree. These
curious phenomena continued " in the highest
splendour" till past eight o' clock, although a black
cloud had for some time totally obscured the face
of the sun.

A phenomenon no less marvellous was seen by
Professor Vince of Cambridge and another gentleman
on the 6th August 1806 at Ramsgate. The summits $v\,w\,x\,y$ of the four turrets of Dover Castle are
usually seen over the hill A B, upon which it stands,
lying between Ramsgate and Dover; but on the
day above-mentioned, at seven o'clock in the evening, when the air was very still and a little hazy, not
only were the tops $v\,w\,x\,y$ of the four towers of
Dover Castle seen over the adjacent hill A B, *but the
whole of the Castle m n r s, appeared as if it were
situated on the side of the hill next Ramsgate,*
and rising above the hill as much as usual. This
phenomenon was so very singular and unexpected,

that at first sight Dr Vince thought it an illusion ; but upon continuing his observations, he became satisfied that it was a real image of the Castle.

Fig. 31.

Upon this he gave a telescope to a person present, who, upon attentive examination, saw also a very clear image of the castle, as the Doctor had described it. He continued to observe it for about twenty minutes, during which time the appearance remained precisely the same, but rain coming on, they were prevented from making any farther observations. Between the observers and the land from which the hill rises, there was about six miles of sea, and from thence to the top of the hill there was about the same distance. Their own height above the surface of the water was about seventy feet.

This illusion derived great force from the remarkable circumstance, that the hill itself did not appear through the image, as it might have been expected to do. The image of the castle was very strong and well defined, and though the rays from the hill

behind it must undoubtedly have come to the eye,
yet the strength of the image of the castle so far
obscured the back ground, that it made no sensible
impression on the observers.   Their attention was,
of course, principally directed to the image of the
castle ; but if the hill behind had been at all visible,
Dr Vince conceives that it could not have escaped
their observation, as they continued to look at it
for a considerable time with a good telescope

Hitherto our aërial visions have been seen only
in their erect and natural positions, either projected against the ground or elevated in the air ; but
cases have occurred in which both erect and inverted images of objects have been seen in the air,
sometimes singly, sometimes combined, sometimes
when the real object was invisible, and sometimes
when a part of it had begun to show itself to the
spectator.

In the year 1793, Mr Huddart, when residing at

Fig. 32.

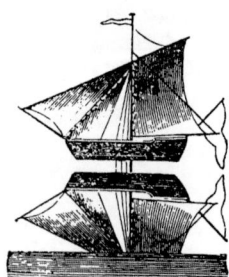

Allonby, in Cumberland, perceived the inverted image
of a ship beneath the image, as shown in Fig. 32 ;

but Dr Vince, who afterwards observed this phenomenon under a greater variety of forms, found that
the ship which was here considered the real one,
was only an erect image of the real ship, which
was at the time beneath the horizon, and wholly invisible.

In August 1798, Dr Vince observed a great variety of these aërial images of vessels approaching
the horizon. Sometimes there was seen only one

Fig. 33.

inverted image above the real ship, and this was
generally the case when the real ship was full in

view. But when the real ship was just beginning
to show its top-mast above the horizon, as at A, Fig.
33, two aërial images of it were seen, one at B inverted, and the other in its natural position at C.
In this case the sea was distinctly visible between
the erect and inverted images, but in other cases
the hull of the one image was immediately in contact
with the hull of the other.

Analogous phenomena were seen by Captain
Scoresby when navigating with the ship Baffin
the icy sea in the immediate neighbourhood of west
Greenland. On the 28th of June 1820, he observed
about eighteen sail of ships at the distance of ten
or fifteen miles. The sun had shone during the
day without the interposition of a cloud, and his
rays were peculiarly powerful. The intensity of
its light occasioned a painful sensation in the eyes,
while its heat softened the tar in the rigging of the
ship, and melted the snow on the surrounding ice
with such rapidity, that pools of fresh water were
formed on almost every place, and thousands of rills
carried the excess into the sea. There was scarcely
a breath of wind : The sea was as smooth as a mirror. The surrounding ice was crowded together,
and exhibited every variety, from the smallest lumps
to the most magnificent sheets. Bears traversed
the fields and floes in unusual numbers, and many
whales sported in the recesses and openings among
the drift ice. About six in the evening, a light
breeze at N. W. having sprung up, a thin stratus
or " fog bank," at first considerably illuminated by
the sun, appeared in the same quarter, and gradually
rose to the altitude of about a quarter of a degree.
At this time most of the ships navigating at the dis-

tance of ten or fifteen miles began to change their form and magnitude, and when examined by a telescope from the mast-head exhibited some extraordinary appearances, which differed at almost every point of the compass. One ship had a perfect image, as dark and distinct as the original, united to its mast head in a reverse position. Two others presented two distinct inverted images in the air, one of them a perfect figure of the original, and the other wanting the hull. Two or three more were strangely distorted, their masts appearing of at least twice their proper height, the top-gallant mast forming one-half of the total elevation, and other vessels exhibited an appearance totally different from all the preceding, being as it were compressed in place of elongated. Their masts seemed to be scarcely one-half of their proper altitude, in consequence of which one would have supposed that they were greatly heeled to one side, or in the position called careening. Along with all the images of the ships a reflexion of the ice, sometimes in two strata, also appeared in the air, and these reflexions suggested the idea of cliffs composed of vertical columns of alabaster.

On the 15th, 16th, and 17th, of the same month, Mr Scoresby observed similar phenomena, sometimes extending continuously through half the circumference of the horizon, and at other times appearing only in detached spots in various quarters. The inverted images of distant vessels were often seen in the air, *while the ships themselves were far beyond the reach of vision.* Some ships were elevated to twice their proper height, while others were compressed almost to a line. Hummocks of ice were sur-

prisingly enlarged, and every prominent object in a
proper position was either magnified or distorted.

But of all the phenomena witnessed by Mr
Scoresby, that of the *Enchanted Coast*, as it may be
called, must have been the most remarkable, This
singular effect was seen on the 18th July, when
the sky was clear, and a tremulous and perfectly
transparent vapour was particularly sensible and
profuse : At nine o'clock in the morning, when the
phenomenon was first seen, the thermometer stood
at 42° Fahr. but in the preceding evening it
must have been greatly lower, as the sea was in
many places covered with a considerable pellicle of
new ice,—a circumstance which in the very warmest
time of the year must be considered as quite extraordinary, especially when it is known that 10°
farther to the north no freezing of the sea at this
season had ever before been observed. Having approached on this occasion so near the unexplored
shore of Greenland that the land appeared distinct
and bold, Mr Scoresby was anxious to obtain a drawing of it, but on making the attempt he found that
the outline was constantly changing, and he was induced to examine the coast with a telescope, and
to sketch the various appearances which presented
themselves. These are shown, without any regard
to their proper order, in Fig. 34, which we shall
describe in Mr Scoresby's own words : " The general telescopic appearance of the coast was that of an
extensive ancient city abounding with the ruins of
castles, obelisks, churches, and monuments, with
other large and conspicuous buildings. Some of
the hills seemed to be surmounted by turrets, battle-
ments, spires, and pinnacles ; while others, subject-

ed to one or two reflexions, exhibited large masses
of rock, apparently suspended in the air, at a considerable elevation above the actual termination of the

Fig. 34.

mountains to which they referred. The whole
exhibition was a grand phantasmagoria. Scarcely
was any particular portion sketched before it
changed its appearance, and assumed the form of
an object totally different. It was perhaps alternately a castle, a cathedral, or an obelisk ; then
expanding horizontally, and coalescing with the
adjoining hills, united the intermediate vallies,
though some miles in width, by a bridge of a single
arch, of the most magnificent appearance and extent. Notwithstanding these repeated changes, the
various figures represented in the drawing had all
the distinctness of reality ; and not only the different strata, but also the veins of the rocks, with

the wreaths of snow occupying ravines and fissures, formed sharp and distinct lines, and exhibited every appearance of the most perfect solidity."

One of the most remarkable facts respecting aërial images presented itself to Mr Scoresby in a later voyage which he performed to the coast of Greenland in 1822. Having seen an inverted image of a ship in the air he directed to it his telescope ; he was able to discover it to be his father's ship, which was at the time below the horizon. " It was," says he, " so well defined, that I could distinguish by a telescope every sail, the general rig of the ship, and its particular character ; insomuch, that I confidently pronounced it to be my father's ship, the *Fame*, which it afterwards proved to be ; though, on comparing notes with my father, I found that our relative position, at the time, gave a distance from one another very nearly 30 miles, being about seventeen miles beyond the horizon, and some leagues beyond the limit of direct vision. I was so struck with the peculiarity of the circumstance, that I mentioned it to the officer of the watch, stating my full conviction that the Fame was then cruising in the neighbouring inlet."

Several curious effects of the mirage were observed by Baron Humboldt during his travels in South America. When he was residing at Cumana, he frequently saw the islands of Picuita and Boracha suspended in the air, and sometimes with an inverted image. On one occasion he observed small fishing-boats swimming in the air, during more than three or four minutes, above the well defined horizon of the sea, and when they were viewed through a telescope, one of the boats had an inverted image accompany-

ing it in its movements.   This distinguished traveller observed similar phenomena in the barren
steppes of the Caraccas, and on the borders of the
Orinoco, where the river is surrounded by sandy
plains.   Little hills and chains of hills appeared
suspended in the air, when seen from the steppes,
at three or four leagues distance.   Palm trees standing single in the Llanos appeared to be cut off at
bottom, as if a stratum of air separated them from
the ground; and, as in the African desert, plains des-
titute of vegetation appeared to be rivers or lakes.
At the Mesa de Pavona M. Humboldt and M. Bon-
pland *saw cows suspended in the air* at the distance
of 1000 toises, and having their feet elevated 3′ 20″
above the soil.   In this case the images were
erect, but the travellers learned from good authority that *inverted images of horses had been seen suspended in the air* near Calabozo.

   In all these cases of aërial spectres the images
were directly above the real object; but a curious
case was observed by MM. Jurine and Soret on
the 17th September 1818, where the image of the
vessel was on one side of the real one.   About 10^h
P. M. a bark at the distance of about 4000 toises
from Bellerive, on the lake of Geneva, was seen
approaching to Geneva by the *left* bank of the
lake, and at the same time an image of the sails
was observed above the water, which, instead of
following the direction of the bark, separated from
it, and appeared to approach Geneva by the right
bank of the lake, the *image* moving from *east*
to *west*, while the *bark* moved from *north* to
south.   When the image first separated from the
bark they had both the same magnitude, but the

K

image diminished as it receded from it, and was reduced to one-half when the phenomenon disappeared.

A very unusual example of aërial spectres occurred to Dr A. P. Buchan while walking on the cliff about a mile to the east of Brighton on the morning of the 28th November 1804. "While watching the rising of the sun," says he, " I turned my eyes directly towards the sea, just as the solar disk emerged from the surface of the water, and saw the face of the cliff on which I was standing, represented precisely opposite to me at some distance on the ocean. Calling the attention of my companion to this appearance, we discerned our own figures standing on the summit of the apparent opposite cliff, as well as the representation of the windmill near at hand.

"The reflected images were most distinctly precisely opposite to where we stood, and the false cliff seemed to fade away, and to draw near to the real one, in proportion as it receded towards the west. This phenomenon lasted about ten minutes, or till the sun had risen nearly his own diameter above the surface of the ocean. The whole then seemed to be elevated into the air, and successively disappeared, giving an impression very similar to that which is produced by the drawing up of a drop scene in the theatre. The horizon was cloudy, or perhaps it might with more propriety be said that the surface of the sea was covered with a dense fog of many yards in height, and which gradually receded before the rays of the sun."

An illusion of a different kind, though not less interesting, is described by the Reverend Mr Hughes in his Travels in Greece, as seen from the

summit of Mount Ætna. " I must not forget to
mention," says he, "one extraordinary phenomenon,
which we observed, and for which I have searched
in vain for a satisfactory solution. At the extremity of the vast shadow which Ætna projects across
the island, appeared a perfect and distinct image of
the mountain itself elevated above the horizon, and
diminished as if viewed in a concave mirror. Where
or what the reflector could be which exhibited this
image, I cannot conceive ; we could not be mistaken
in its appearance, for all our party observed it, and
we had been prepared for it beforehand by our Cata-
nian friends. It remained visible about *ten* minutes,
and disappeared as the shadow decreased. Mr Jones
observed the same phenomenon, as well as some
other friends with whom I conversed upon the subject in England."

It is impossible to study the preceding phenomena without being impressed with the conviction,
that nature is full of the marvellous, and that the
progress of science and the diffusion of knowledge
are alone capable of dispelling the fears which her
wonders must necessarily excite even in enlightened
minds. When a spectre haunts the couch of the
sick, or follows the susceptible vision of the invalid, a consciousness of indisposition divests the apparition of much of its terror, while its invisibility
to surrounding friends soon stamps it with the impress of a false perception. The spectres of the
conjuror too, however skilfully they may be raised,
quickly lose their supernatural character, and even
the most ignorant beholder regards the modern magician as but an ordinary man, who borrows from
the sciences the best working implements of his

art. But when, in the midst of solitude, and in
situations where the mind is undisturbed by sub-
lunary cares, we see our own image delineated in
the air, and mimicking in gigantic perspective the tiny
movements of humanity ;—when we see troops in
military array performing their evolutions on the very
face of an almost inaccessible precipice ;—when in
the eye of day a mountain seems to become transparent, and exhibits on one side of it a castle which
we know to exist only on the other ;—when distant
objects concealed by the roundness of the earth,
and beyond the cognizance of the telescope, are actually transferred over the intervening convexity and
presented in distinct and magnified outline to our accurate examination ;—when such varied and striking phantasms are seen also by all around us, and
therefore appear in the character of real phenomena of nature, our impressions of supernatural agency can only be removed by a distinct and satisfactory knowledge of the causes which gave them birth.

It is only within the last forty years that science
has brought these atmospherical spectres within
the circle of her dominion; and not only are all
their phenomena susceptible of distinct explanation,
but we can even reproduce them on a small scale
with the simplest elements of our optical apparatus.

In order to convey a general idea of the causes
of these phenomena, let A B C D, Fig. 35, be a glass
trough filled with water, and let a small ship be
placed at S.   An eye situated about E, will see the
top-mast of the ship S, directly through the plate
of glass B D.   Fix a convex lens a of short focus
upon the plate of glass B D, and a little above a
straight line S E joining the ship and the eye ; and

immediately above the convex lens *a* place a concave one *b*.   The eye will now see through the

Fig. 35.

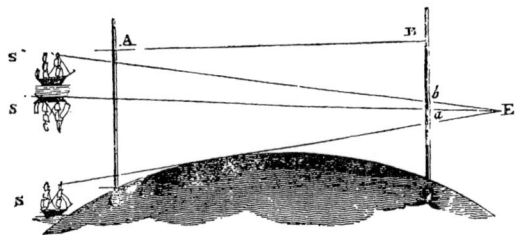

convex lens *a* an *inverted* image of the ship at S',
and through the concave lens *b*, an erect image of
the ship at S'', representing in a general way the
phenomena shown in Fig. 33.   But it will be asked, where are the lenses in nature to produce these
effects?   This question is easily answered.   If we
take a tin tube with glass plates at each end, and
fill it with water, and if we cool it on the outside
with ice, it will act like a *concave* lens when the
cooling effect has reached the axis; and, on the other
hand, if we heat the same tube filled with water,
on the outside, it will act as a *convex* glass.   In
the first case the density of the water diminishes
towards the centre, and in the second it increases towards the centre.   The very same effects are produced in the air, only a greater tract of air is necessary for showing the effect produced, by heating
and cooling it unequally.   If we now remove the
lenses *a, b,* and hold a heated iron horizontally

above the water in the trough A B C, the heat will gradually descend, expanding or rendering rarer the upper portions of the fluid. If, when the heat has reached within a little of the bottom, we look through the trough at the ship S in the direction E S', we shall see an inverted image at S', and an erect one at S", and if we hide from the eye at E all the ship S excepting the top-mast, we shall have an exact representation of the phenomenon in Fig. 33. The experiment will succeed better with oil in place of water ; and the same result may be obtained without heat, by pouring clear syrup into the glass trough till it is nearly one-third full, and then filling it up with water. The water will gradually incorporate with the syrup, and produce, as Dr Wollaston has shown, a regular gradation of density, diminishing from that of the pure syrup to that of the pure water. Similar effects may be obtained by using masses of transparent solids, such as glass, rock salt, &c.

Now it is easy to conceive how the changes of density which we can thus produce artificially may be produced in nature. If in serene weather the surface of the sea is much colder than the air of the atmosphere, as it frequently is, and as it was to a very great degree during the phenomena described by Mr Scoresby, the air next the sea will gradually become colder and colder, by giving out its heat to the water; and the air immediately above will give out its heat to the cooler air immediately below it, so that the air from the surface of the sea, to a considerable height upwards, will gradually diminish in density, and therefore must produce the very phenomena we have described.

The phenomenon of Dover Castle seen on the Ramsgate side of the hill was produced by the air being more dense near the ground, and above the sea, than at greater heights, and hence the rays proceeding from the castle reached the eye in curve lines, and the cause of its occupying its natural position on the hill, and not being seen in the air, was that the top of the hill itself, in consequence of being so near the castle, suffered the same change from the varying density of the air, and therefore, the castle and the hill were equally elevated and retained their relative positions. The reason why the image of the castle and the hill appeared erect was, that the rays from the top and bottom of the castle had not crossed before they reached Ramsgate; but as they met at Ramsgate, an eye at a greater distance from the castle, and in the path of the rays, would have seen the image inverted. This will be

Fig. 36.

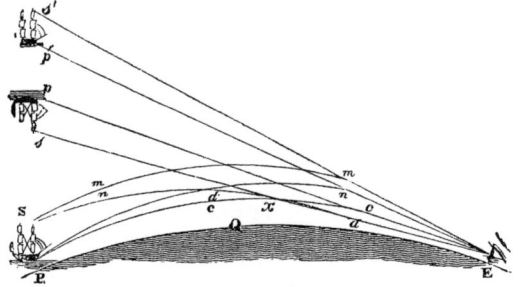

better understood from the annexed diagram, which represents the actual progress of the rays, from a ship S P, concealed from the observer at E by the

convexity of the earth P Q E. A ray proceeding
from the keel of the ship P is refracted into the
curve line P c x c E, and a ray proceeding from the
top-mast S, is refracted in the direction S d x d E,
the two rays crossing at x, and proceeding to the
eye E with the ray from the keel P uppermost;
hence the ship must appear inverted as at s p. Now
if the eye E of the observer had been placed nearer the ship as at x, before the rays crossed, as
was the case at Ramsgate, it would have seen an
erect image of the ship raised a little above the real
ship S P. Rays S m, S n, proceeding higher up
in the air, are refracted in the directions S m m E,
S n n E, but do not cross before they reach the eye,
and therefore they afford the erect image of the
ship shown at s' p'.

The aërial troopers seen at Souterfell were produced by the very same process as the spectre of
Dover Castle, having been brought by unequal refraction from one side of the hill to the other. It
is not our business to discover how a troop of sol-
diers came to be performing their evolutions on
the other side of Souterfell; but if there was then
no road along which they could be marching, it is
highly probable that they were troops exercising
among the hills in secret previous to the breaking
out of the rebellion in 1745.

The image of the Genevese bark which was seen
sailing at a distance from the real one, arose from
the same cause as the images of ships in the air,
with this difference only, that in this case the strata
of equal density were vertical or perpendicular to
the water, whereas in the former cases they were
horizontal or parallel to the water. The state of

the air which produced the lateral image may be produced by a headland or island, or even rocks, near the surface, and covered with water.   These headlands, islands, or sunken rocks being powerfully heated by the sun in the day-time, will heat the air immediately above them, while the adjacent air over the sea will retain its former coolness and density.   Hence there will necessarily arise a gradation of density varying in the same horizontal direction, or where the lines of equal density are vertical.   If we suppose the very same state of the air to exist in a horizontal plane which exists in a vertical plane, in Fig. 36, then the same images would be seen in a horizontal line, viz. an inverted one at $s\,p$, and an erect one at $s'\,p'$. In the case of the Genevese bark the rays had not crossed before they reached the eye, and therefore the image was an erect one.   Had the real Genevese bark been concealed by some promontory or other cause from the observation of M.M. Jurine and Soret, they might have attached a supernatural character to the spectral image, especially if they had seen it gradually decay, and finally disappear on the still and unbroken surface of the lake.   No similar fact had been previously observed, and there were no circumstances in the case to have excited the suspicion, that it was the spectre of a real vessel produced by unequal refraction.

The spectre of the Brocken and other phenomena of the same kind have essentially a different origin from those which arise from unequal refraction. They are merely shadows of the observer projected on dense vapour or thin fleecy clouds, which have the power of reflecting much light.   They are

seen most frequently at sunrise, because it is at that
time that the vapours and clouds necessary for their
production are most likely to be generated ; and
they can be seen only when the sun is throwing
his rays horizontally, because the shadow of the
observer would otherwise be thrown either up in
the air, or down upon the ground. If there are
two persons looking at the phenomenon, as when
M. Haue and the landlord saw it together, each
observer will see his own image most distinctly,
and the head will be more distinct than the rest
of the figure, because the rays of the sun will be
more copiously reflected at a perpendicular incidence ; and as from this cause, the light reflected
from the vapour or cloud becomes fainter farther
from the shadow, the appearance of a halo round
the head of the observer is frequently visible. M.
Haue mentions the extraordinary circumstance of
the two spectres of him and the landlord being
joined by a *third figure,* but he unfortunately does
not inform us which of the two figures was doubled, for it is impossible that a person could have
joined their party unobserved. It is very probable
that the new spectre forms a natural addition to
the group, as we have represented it in Fig. 30,
and if this was the case, it could only have been
produced by a duplication of one of the figures
produced by unequal refraction.

The reflected spectre of Dr Buchan standing
upon the cliff at Brighton arose from a cause to
which we have not yet adverted. It was obviously
no shadow, for it is certain from the locality,
that the rays of the sun fell upon the face of the
cliff and upon his person at angle of about 73°

from the perpendicular, so as to illuminate them strongly. Now there are two ways in which such an image may have been reflected, namely, either from strata of air of variable density, or from a vertical stratum of vapour, consisting of exceedingly minute globules of water. Whenever light suffers refraction, either in passing at once from one medium into another, or from one part of the same medium into another of different density, a portion of it suffers reflexion. If an object, therefore, were strongly illuminated, a sufficiently distinct image, or rather shadow of it, might be seen by reflexion from strata of air of different density. As the temperature at which moisture is deposited in the atmosphere varies with the density of the air, then at the same temperature moisture might be depositing in a stratum of one density, while no deposition is taking place in the adjacent stratum of a different density. Hence there would exist as it were in the air a vertical wall or stratum of minute globules of water, from the surface of which a sufficiently distinct image of a highly illuminated object might be reflected. That this is possible may be proved by breathing upon glass. If the particles deposited upon the glass are large, then no distinct reflexion will take place; but if the particles be very small, we shall see a distinct image formed by the surface of the aqueous film.

The phenomena of the Fata Morgana have been too imperfectly described to enable us to offer a satisfactory explanation of them. The aërial images are obviously those formed by unequal refraction. The pictures seen on the sea may be either the aërial images reflected from its surface,

or from a stratum of dense vapour, or they may
be the direct reflexions from the objects themselves.   The coloured images, as described by
Minasi, have never been seen in any analogous
phenomena, and require to be better described
before they can be submitted to scientific examination.

The representation of ships in the air by unequal refraction has no doubt given rise in early
times to those superstitions which have prevailed
in different countries respecting " phantom ships"
as Mr Washington Irving calls them, which always sail in the eye of the wind, and plough their
way through the smooth sea, where there is not a
breath of wind upon its surface.   In his beautiful
story of the storm ship, which makes its way up
the Hudson against wind and tide, this elegant
writer has finely embodied one of the most interesting superstitions of the early American colo-
nists.   The Flying Dutchman had in all probability a similar origin, and the wizard beacon-keeper
of the Isle of France, who saw in the air the vessels bound to the island long before they appeared
in the offing, must have derived his power from a
diligent observation of the phenomena of nature.

LETTER VII.

Illusions depending on the ear—Practised by the ancients—
Speaking and singing heads of the ancients—Exhibition
of the invisible girl described and explained—Illusions
arising from the difficulty of determining the direction of
sounds—Singular example of this illusion—Nature of ven-
triloquism—Exhibitions of some of the most celebrated
ventriloquists—M. St Gille—Louis Brabant—M. Alex-
andre—Capt. Lyon's account of Eskimaux ventriloquists.

NEXT to the eye the ear is the most fertile source
of our illusions, and the ancient magicians seem
to have been very successful in turning to their
purposes the doctrines of sound.   In the labyrinth of Egypt, which contained twelve palaces
and 1500 subterraneous apartments, the gods were
made to speak in a voice of thunder; and Pliny,
in whose time this singular structure existed, informs
us, that some of the palaces were so constructed
that their doors could not be opened without per-
mitting the peals of thunder from being heard in
the interior.   When Darius Hystaspes ascended
the throne, and allowed his subjects to prostrate
themselves before him as a god, the divinity of his
character was impressed upon his worshippers by
the bursts of thunder and flashes of lightning which
accompanied their devotion.   History has of course
not informed us how these effects were produced;

but it is probable that, in the subterraneous and
vaulted apartments of the Egyptian labyrinth, the
reverberated sounds arising from the mere opening
and shutting of the doors themselves afforded a
sufficient imitation of ordinary thunder. In the
palace of the Persian king, however, a more artificial imitation is likely to have been employed, and
it is not improbable that the method used in our
modern theatres was known to the ancients. A
thin sheet of iron, three or four feet long, such as
that used for German stoves, is held by one corner
between the finger and the thumb, and allowed to
hang freely by its own weight. The hand is then
moved or shaken horizontally, so as to agitate the
corner in a direction at right angles to the surface of the sheet. By this simple process a great
variety of sounds will be produced, varying from
the deep growl of distant thunder to those loud
and explosive bursts which rattle in quick succession from clouds immediately over our heads.
The operator soon acquires great power over this
instrument, so as to be able to produce from it any
intensity and character of sound that may be required.
The same effect may be produced by sheets of tin-
plate, and by thin plates of mica; but on account
of their small size, the sound is shorter and more
acute. In modern exhibitions an admirable imitation of lightning is produced by throwing the powder of rosin, or the dust of lycopodium, through a
flame, and the rattling showers of rain which accompany these meteors are well imitated by a well
regulated shower of peas.

The principal pieces of acoustic mechanism used
by the ancients were *speaking* or *singing heads*,

which were constructed for the purpose of representing the gods, or of uttering oracular responses.
Among these, the speaking head of Orpheus,
which uttered its responses at Lesbos, is one of the
most famous.   It was celebrated not only throughout Greece, but even in Persia, and it had the credit of predicting, in the equivocal language of the
heathen oracles, the bloody death which terminated the expedition of Cyrus the Great into Scythia.
Odin, the mighty magician of the north, who im-
ported into Scandinavia the magical arts of the east,
possessed a speaking head, said to be that of the
sage Minos, which he had enchased in gold, and
which uttered responses that had all the authority
of a divine revelation.   The celebrated mechanic
Gerbert, who filled the papal chair A. D. 1000,
under the name of Sylvester II., constructed a
speaking head of brass.   Albertus Magnus is said
to have executed a head in the thirteenth century,
which not only moved but spoke.   It was made of
earthen-ware, and Thomas Aquinas is said to have
been so terrified when he saw it, that he broke it
in pieces, upon which the mechanist exclaimed,
" There goes the labour of thirty years."

It has been supposed by some authors, that in
the ancient speaking machines the deception is effected by means of ventriloquism, the voice issuing
from the juggler himself; but it is more probable
that the sound was conveyed by pipes from a person in another apartment to the mouth of the
figure.   Lucian, indeed, expressly informs us, that
the impostor Alexander made his figure of Æscula-
pius speak, by transmitting his voice through the
gullet of a crane to the mouth of the statue ; and

that this method was general, appears from a passage in Theodoretus, who assures us, that in the fourth century, when Bishop Theophilus broke to pieces the statues at Alexandria, he found some which were hollow, and which were so placed against a wall, that the priest could conceal himself behind them, and address the ignorant spectators through their mouths.

Even in modern times, speaking-machines have been constructed on this principle. The figure is frequently a mere head placed upon a hollow pedestal, which, in order to promote the deception, contains a pair of bellows, a sounding-board, a cylinder and pipes supposed to represent the organs of speech. In other cases these are dispensed with, and a simple wooden head utters its sounds through a speaking trumpet. At the court of Charles II. this deception was exhibited with great effect by one Thomas Irson, an Englishman, and when the astonishment had become very general, a popish priest was discovered by one of the pages in an adjoining apartment. The questions had been proposed to the wooden figure by whispering into its ear, and this learned personage had answered them all with great ability, by speaking through a pipe in the same language in which the questions were proposed. Professor Beckmann informs us that children and women were generally concealed either in the juggler's box, or in the adjacent apartment, and that the juggler gave them every assistance by means of signs previously agreed upon. When one of these exhibitions was shown at Gottingen, the Professor was allowed, on the promise of secrecy, to witness the process of deception. He saw the assistant in another room, standing before the pipe with a card in his hand, upon which the signs agreed upon had been marked, and he had been introduced so privately into the house that even the landlady was ignorant of his being there.

An exhibition of the very same kind has been brought forward in our own day, under the name of the *Invisible Girl;* and as the mechanism employed was extremely ingenious, and is well fitted to convey an idea of this class of deceptions, we shall give a detailed description of it.

The machinery, as constructed by M. Charles, is shown in Fig. 37 in perspective, and a plan of it

Fig. 37.

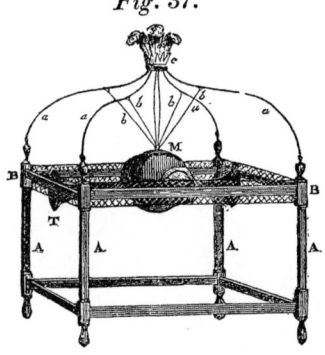

in Fig. 38. The four upright posts A, A, A, A, are united at top by a cross rail B, B, and by two similar rails at bottom. Four bent wires *a, a, a, a,* proceeded from the top of these posts, and terminated at *c.* A hollow copper ball M, about a foot in diameter, was suspended from these wires by four

L

slender ribbands *b, b, b, b,* and into the copper ball
were fixed the extremities of four trumpets T, T,
T, T, with their mouths outwards.

Fig. 38.

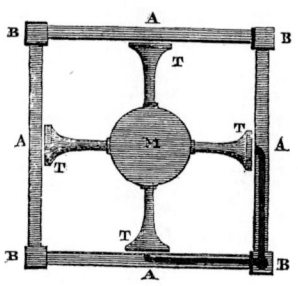

The apparatus now described was all that was
visible to the spectator; and though fixed in one
spot, yet it had the appearance of a piece of separate machinery, which might have occupied any other
part of the room. When one of the spectators was
requested by the exhibitor to propose some question, he did it by speaking into one of the trumpets
at T.   An appropriate answer was then returned
from all the trumpets, and the sound issued with
sufficient intensity to be heard by an ear applied to
any of them, and yet it was so weak that it appeared to come from a person of very diminutive size.
Hence the sound was supposed to come from an
invisible girl, though the real speaker was a full
grown woman.   The invisible lady conversed in
different languages, sang beautifully, and made the
most lively and appropriate remarks on the persons in the room.

This exhibition was obviously far more wonderful than the speaking heads which we have described, as the latter invariably communicated with a wall, or with a pedestal through which pipes could be carried into the next apartment. But the ball M and its trumpets communicated with nothing through which sound could be conveyed. The spectator satisfied himself by examination that the ribbands *b, b,* were real ribbands, which concealed nothing, and which could convey no sound, and as he never conceived that the ordinary piece of framework A B, could be of any other use than its apparent one of supporting the sphere M, and defending it from the spectators, he was left in utter amazement respecting the origin of the sound, and his surprise was increased by the difference between the sounds which were uttered and those of ordinary speech.

Though the spectators were thus deceived by their own reasoning, yet the process of deception was a very simple one. In two of the horizontal railings A, A, Fig. 38, opposite the trumpet mouths T, there was an aperture communicating with a pipe or tube which went to the vertical post B, and descending it, as shown at T A A, Fig. 39, went beneath the floor *f f* in the direction *p, p,* and entered the apartment N, where the invisible lady sat. On the side of the partition about *h,* there was a small hole, through which the lady saw what was going on in the exhibition room, and communications were no doubt made to her by signals from the person who attended the machine. When one of the spectators asked a question by speaking into one of the trumpets T, the sound was re-

flected from the mouth of the trumpet back to the
aperture at A, in the horizontal rail, Fig. 38, and

Fig. 39.

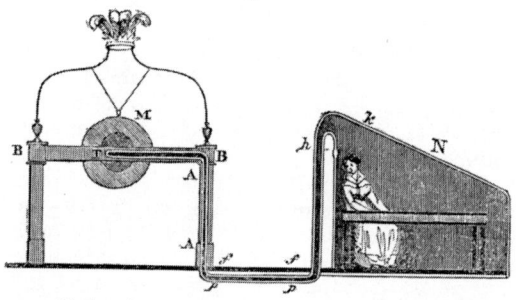

was distinctly conveyed along the closed tube into the apartment N.   In like manner the answer
issued from the aperture A, and being reflected
back to the ear of the spectator by the trumpet,
he heard the sounds with that change of character
which they receive when transmitted through a
tube and then reflected to the ear.

The surprise of the auditors was greatly increased by the circumstance, that an answer was returned to questions put in a whisper, and also by the
conviction that nobody but a person in the middle
of the audience could observe the circumstances to
which the invisible figure frequently adverted.

Although the performances of speaking heads
were generally effected by the methods now described, yet there is reason to think that the ventriloquist sometimes presided at the exhibition, and
deceived the audience by his extraordinary powers
of illusion.   There is no species of deception more

irresistible in its effects than that which arises from
the uncertainty with which we judge of the direction and distance of sounds. Every person must
have noticed how a sound in their own ears is often mistaken for some loud noise moderated by the
distance from which it is supposed to come ; and
the sportsman must have frequently been surprised
at the existence of musical sounds humming re-
motely in the extended heath, when it was only
the wind sounding in the barrel of his gun. The
great proportion of apparitions that haunt old castles
and apartments associated with death exist only in
the sounds which accompany them. The imagination even of the boldest inmate of a place hallowed by superstition, will transfer some trifling
sound near his own person to a direction and to a
distance very different from the truth, and the sound
which otherwise might have no peculiar com-
plexion will derive another character from its new
locality. Spurning the idea of a supernatural origin, he determines to unmask the spectre, and grap-
ple with it in its den. All the inmates of the
house are found to be asleep—even the quadrupeds
are in their lair—there is not a breath of wind to
ruffle the lake that reflects through the casement,
the reclining crescent of the night ; and the mas-
sive walls in which he is inclosed forbid the idea
that he has been disturbed by the warping of pan-
nelling or the bending of partitions. His search
is vain ; and he remains master of his own secret,
till he has another opportunity of investigation.
The same sound again disturbs him, and, modified
probably by his own position at the time, it may
perhaps appear to come in a direction slightly dif-

ferent from the last. His searches are resumed, and he is again disappointed. If this incident should recur night after night with the same result ;—if the sound should appear to depend upon his own motions, or be any how associated with himself, with his present feelings, or with his past history, his personal courage will give way, a superstitious dread, at which he himself perhaps laughs, will seize his mind, and he will rather believe that the sounds have a supernatural origin, than that they could continue to issue from a spot where he knows there is no natural cause for their production.

I have had occasion to have personal knowledge of a case much stronger than that which has now been put. A gentleman, devoid of all superstitious feelings, and living in a house free from any gloomy associations, heard night after night in his bed-room a singular noise, unlike any ordinary sound to which he was accustomed. He had slept in the same room for years without hearing it, and he attributed it at first to some change of circumstances in the roof or in the walls of the room, but after the strictest examination no cause could be found for it. It occurred only once in the night, it was heard almost every night, with few interruptions. It was over in an instant, and it never took place till after the gentleman had gone to bed. It was always distinctly heard by his companion, to whose time of going to bed it had no relation. It depended on the gentleman alone, and it followed him into another apartment with another bed, on the opposite side of the house. Accustomed to such investigations, he made the most diligent but fruitless search into its cause. The consideration that the sound had a special reference to him alone, operated upon his imagination, and he did not scruple to acknowledge that the recurrence of the mysterious sound produced a superstitious feeling at the moment. Many months afterwards it was found that the sound arose from the partial opening of the door of a wardrobe which was within a few feet of the gentleman's head, and which had been taken into the other apartment. This wardrobe was almost always opened before he retired to bed, and the door being a little too tight, it gradually forced itself open with a sort of dull sound, resembling the note of a drum.   As the door had only started half an inch out of its place, its change of place never attracted attention. The sound, indeed, seemed to come in a different direction, and from a greater distance.

When sounds so mysterious in their origin are heard by persons predisposed to a belief in the marvellous, their influence over the mind must be very powerful.   An inquiry into their origin, if it is made at all, will be made more in the hope of confirming than of removing the original impression, and the unfortunate victim of his own fears will also be the willing dupe of his own judgment.

This uncertainty with respect to the direction of sound is the foundation of the art of ventriloquism. If we place ten men in a row at such a distance from us that they are included in the angle within which we cannot judge of the direction of sound, and if in a calm day each of them speaks in succession, we shall not be able with closed eyes to determine from which of the ten men any of the sounds proceeds, and we shall be incapable of perceiving that there is any difference in the direction of the sounds emitted by the two outermost : If a man and a child are placed within the same angle, and if the man speaks with the accent of a child without any corresponding motion in his mouth or face, we shall necessarily believe that the voice comes from the child : Nay, if the child is so distant from the man that the voice actually appears to us to come from the man, we will still continue in the belief that the child is the speaker; and this conviction would acquire additional strength if the child favoured the deception by accommodating its features and gestures to the words spoken by the man. So powerful, indeed, is the influence of this deception, that if a jack-ass placed near the man were to open its mouth, and shake its head responsive to the words uttered by his neighbour, we would rather believe that the ass spoke than that the sounds proceeded from a person whose mouth was shut, and the muscles of whose face were in perfect repose. If our imagination were even directed to a marble statue or a lump of inanimate matter, as the source from which we were to expect the sounds to issue, we would still be deceived, and would refer the sounds even to these lifeless objects. The illusion would be greatly promoted if the voice were totally different in its tone and character from that of the man from whom it really comes; and if he occasionally speak in his own full and measured voice, the belief will be irresistible that the assumed voice proceeds from the quadruped or from the inanimate object.

When the sounds which are required to proceed from any given object are such as they are actually calculated to yield, the process of deception is extremely easy, and it may be successfully executed even if the angle between the real and the supposed directions of the sound is much greater than the angle of uncertainty. Mr Dugald Stewart has stated some cases in which deceptions of this kind were very perfect. He mentions his having seen a person who, by counterfeiting the gesticulations of a performer on the violin, while he imitated the music by his voice, rivetted the eyes of his audience on the instrument, though every sound they heard proceeded from his own mouth. The late Savile Carey, who imitated the whistling of the wind through a narrow chink, told Mr Stewart that he had frequently practised this deception in the corner of a coffee-house, and that he seldom failed to see some of the company rise to examine the tightness of the windows, while others, more intent on their newspapers, contented themselves with putting on their hats and buttoning their coats. Mr Stewart likewise mentions an exhibition formerly common in some of the continental theatres, where a performer on the stage displayed the dumb show of singing with his lips and eyes and gestures, while another person unseen supplied the music with his voice. The deception in this case he found to be at first so complete as to impose upon the nicest ear and the quickest eye ; but in the progress of the entertainment, he came distinctly sensible of the imposition, and sometimes wondered that it should have misled him for a moment. In this case there can be no doubt that the deception was at first the work of the imagination, and was not sustained by the acoustic principle. The real and the mock singer were too distant, and when the influence of the imagination subsided, the true direction of the sound was discovered. This detection of the imposture, however, may have arisen from another cause. If the mock singer happened to change the position of his head, while the real singer made no corresponding change in his voice, the attentive spectator would at once notice this incongruity, and discover the imposition.

In many of the feats of ventriloquism the performer contrives, under some pretence or other, to conceal his face, but ventriloquists of great distinction, such as M. Alexandre, practise their art without any such concealment.

Ventriloquism loses its distinctive character if its imitations are not performed by a voice from the belly. The voice, indeed, does not actually come from that region, but when the ventriloquist utters sounds from the larynx without moving the muscles of his face, he gives them strength by a powerful action of the abdominal muscles. Hence he speaks by means of his belly, although the throat is the real source from which the sounds proceed. Mr Dugald Stewart has doubted the fact, that ventriloquists possess the power of fetching a voice from within : he cannot conceive what aid could be derived from such an extraordinary power ; and he considers that the imagination, when seconded by such powers of imitation as some mimics possess, is quite sufficient to account for all the phenomena of ventriloquism which he has heard. This opinion, however, is strongly opposed by the remark made to Mr Stewart himself by a ventriloquist, " that his art would be perfect if it were possible only to speak distinctly without any movement of the lips at all." But, independent of this admission, it is a matter of abso-

lute certainty, that this internal power is exercised
by the true ventriloquist. In the account which
the Abbé Chapelle has given of the performances of
M. St Gille and Louis Brabant, he distinctly states that
M. St Gille appeared to be absolutely mute while
he was exercising his art, and that no change in his
countenance could be discovered. * He affirms
also that the countenance of Louis Brabant exhibited no change, and that his lips were close and inac-
tive. M. Richerand, who attentively watched the
performances of M. Fitz-James, assures us that during his exhibition there was a distension in the
epigastric region, and that he could not long continue the exertion without fatigue.

The influence over the human mind which the
ventriloquist derives from the skilful practice of his
art is greater than that which is exercised by any
other species of conjuror. The ordinary magician
requires his theatre, his accomplices, and the instruments of his art, and he enjoys but a local sovereignty within the precincts of his own magic
circle. The ventriloquist, on the contrary, has the
supernatural always at his command. In the open
fields, as well as in the crowded city, in the private
apartment, as well as in the public hall, he can sum-
mon up innumerable spirits; and though the persons
of his fictitious dialogue are not visible to the eye,
yet they are as unequivocally present to the imagination of his auditors as if they had been shadowed forth in the silence of a spectral form. In
order to convey some idea of the influence of this
illusion, I shall mention a few well authenticated
cases of successful ventriloquism.

* Edinburgh Journal of Science, No. xviii. p. 254.

M. St Gille, a grocer of St Germain en Lay, whose performances have been recorded by the Abbé de la Chapelle, had occasion to shelter himself from a storm in a neighbouring convent, where the monks were in deep mourning for a much esteemed member of their community who had been recently buried. While lamenting over the tomb of their deceased brother the slight honours which had been paid to his memory, a voice was suddenly heard to issue from the roof of the choir bewailing the condition of the deceased in purgatory, and reproving the brotherhood for their want of zeal. The tidings of this supernatural event brought the whole brotherhood to the church. The voice from above repeated its lamentations and reproaches, and the whole convent fell upon their faces, and vowed to make a reparation of their error. They accordingly chaunted in full choir a *de profundis,* during the intervals of which the spirit of the departed monk expressed his satisfaction at their pious exercises. The prior afterwards inveighed against modern scepticism on the subject of apparitions, and M. St Gille had great difficulty in convincing the fraternity that the whole was a deception.

On another occasion, a commission of the Royal Academy of Sciences at Paris, attended by several persons of the highest rank, met at St Germain en Lay to witness the performances of M. St Gille. The real object of their meeting was purposely withheld from a lady of the party, who was informed that an aërial spirit had lately established itself in the neighbourhood, and that the object of the assembly was to investigate the matter. When the party had sat down to dinner in the open air, the

spirit addressed the lady in a voice which seemed
to come from above their heads, from the surface
of the ground at a great distance, or from a considerable depth under her feet.   Having been thus
addressed at intervals during two hours the lady
was firmly convinced of the existence of the spirit,
and could with difficulty be undeceived.

Another ventriloquist, Louis Brabant, who had been
valet de chambre to Francis I. turned his powers to
a more profitable account.   Having fallen in love
with a rich and beautiful heiress, he was rejected by
her parents as an unsuitable match for their daughter.   On the death of her father, Louis paid a visit
to the widow, and he had no sooner entered the
house than she heard the voice of her deceased husband addressing her from above, " Give my daughter in marriage to Louis Brabant, who is a man of
large fortune and excellent character.   I endure
the inexpressible torments of purgatory for having
refused her to him.   Obey this admonition, and give
everlasting repose to the soul of your poor husband."
This awful command could not be resisted, and the
widow announced her compliance with it.

As our conjuror, however, required money for the
completion of his marriage, he resolved to work upon the fears of one Cornu, an old banker at Lyons,
who had amassed immense wealth by usury and ex-
tortion.   Having obtained an interview with the
miser, he introduced the subjects of demons and
spectres and the torments of purgatory, and during
an interval of silence, the voice of the miser's deceased father was heard complaining of his dreadful situation in purgatory, and calling upon his son to
rescue him from his sufferings by enabling Louis

Brabant to redeem the Christians that were enslav-
ed by the Turks.   The awe-struck miser was also
threatened with eternal damnation if he did not thus
expiate his own sins ; but such was the grasp that
the banker took of his gold that the ventriloquist
was obliged to pay him another visit.   On this occasion, not only his father but all his deceased rela-
tions appealed to him in behalf of his own soul and
theirs, and such was the loudness of their complaints
that the spirit of the banker was subdued, and he
gave the ventriloquist ten thousand crowns to libe-
rate the Christian captives.   When the miser was
afterwards undeceived, he is said to have been so
mortified that he died of vexation.

The ventriloquists of the nineteenth century
made great additions to their art, and the performances of M. Fitz-James and M. Alexandre, which
must have been seen by many of our countrymen,
were far superior to those of their predecessors.
Besides the art of speaking by the muscles of the
throat and the abdomen, without moving those of
the face, these artists had not only studied with
great diligence and success the modifications which
sounds of all kinds undergo from distance, obstructions and other causes, but had acquired the art of
imitating them in the highest perfection.   The
ventriloquist was therefore able to carry on a dialogue in which the *dramatis voces,* as they may be
called, were numerous ; and when on the outside of
an apartment he could personate a mob with its infinite variety of noise and vociferation.   Their influence over an audience was still farther extended
by a singular power over the muscles of the body.
Mr Fitz-James actually succeeded in making the op-

posite or corresponding muscles act differently from each other ; and while one side of his face was merry and laughing, the other was full of sorrow and in tears. At one moment he was tall, thin, and melancholic, and after passing behind a screen he came out "bloated with obesity and staggering with fulness." M. Alexandre possessed the same power over his face and figure, and so striking was the contrast of two of these forms, that an excellent sculptor, Mr Joseph, has perpetuated them in marble.

This new acquirement of the ventriloquist enabled him, in his own single person and with his own single voice, to represent upon the stage a dramatic composition which would have required the assistance of several actors. Although only one character in the piece could be seen at the same time, yet they all appeared during its performance, and the change of face and figure on the part of the ventriloquist was so perfect that his personal identity could not be recognized in the *dramatis personæ*. This deception was rendered still more complete by a particular construction of the dresses, which enabled the performer to reappear in a new character after an interval so short that the audience necessarily believed that it was another person.

It is a curious circumstance that Captain Lyon found among the Eskimaux of Igloolik ventriloquists of no mean skill. There is much rivalry amongst the professors of the art, who do not expose each other's secrets, and their exhibitions derive great importance from the rarity of their occurrence. The following account of one of them is so interesting that we shall give the whole of it in Captain Lyon's words.

" Amongst our Igloolik acquaintances were two females and a few male wizards, of whom the principal was Toolemak. This personage was cunning and intelligent, and, whether professionally, or from his skill in the chace, but perhaps from both reasons, was considered by all the tribe as a man of importance. As I invariably paid great deference to his opinion on all subjects connected with his calling, he freely communicated to me his superior knowledge, and did not scruple to allow of my being present at his interviews with Tornga, or his patron spirit. In consequence of this, I took an early opportunity of requesting my friend to exhibit his skill in my cabin. His old wife was with him, and by much flattery and an accidental display of a glittering knife and some beads, she assisted me in obtaining my request. All light excluded, our sorcerer began chanting to his wife with great vehemence, and she in return answered by singing the Amna-aya, which was not discontinued during the whole ceremony. As far as I could hear, he afterwards began turning himself rapidly round, and in a loud powerful voice vociferated for Tornga with great impatience, at the same time blowing and snorting like a walrus. His noise, impatience, and agitation, increased every moment, and he at length seated himself on the deck, varying his tones, and making a rustling with his clothes. Suddenly the voice seemed smothered, and was so managed as to sound as if retreating beneath the deck, each moment becoming more distant, and ultimately giving the idea of being many feet below the cabin, when it ceased entirely. His wife now, in answer to my queries, informed me very seriously, that he had dived, and that he would send up Tornga. Accordingly, in about half a minute, a distant blowing was heard very slowly approaching, and a voice, which differed from that at first heard, was at times mingled with the blowing, until at length both sounds became distinct, and the old woman informed me that Tornga was come to answer my questions. I accordingly asked several questions of the sagacious spirit, to each of which inquiries I received an answer by two loud claps on the deck, which I was given to understand were favourable.

" A very hollow, yet powerful voice, certainly much different from the tones of Toolemak, now chanted for some time, and a strange jumble of hisses, groans, shouts, and gabblings like a turkey succeeded in rapid order. The old woman sang with increased energy, and as I took it for granted that this was all intended to astonish the Kabloona, I cried repeatedly that I was very much afraid. This, as I expected, added fuel to the fire, until the poor immortal, exhausted by its own might, asked leave to retire.

" The voice gradually sunk from our hearing as at first, and a very indistinct hissing succeeded; in its advance, it sounded like the tone produced by the wind on the bass chord of an Æolian harp. This was soon changed to a rapid hiss like that of a rocket, and Toolemak with a yell announced his return. I had held my breath at the first distant hissing, and twice exhausted myself, yet our conjuror did not once respire, and even his returning

M

and powerful yell was uttered without a previous
stop or inspiration of air.

" Light being admitted, our wizard, as might be
expected, was in a profuse perspiration, and certainly much exhausted by his exertions, which had
continued for at least half an hour.   We now observed a couple of bunches, each consisting of two
stripes of white deer-skin and a long piece of sinew,
attached to the back of his coat.   These we had
not seen before, and were informed that they had
been sewn on by Tornga while he was below." *

Captain Lyon had the good fortune to witness
another of Toolemak's exhibitions, and he was
much struck with the wonderful steadiness of the
wizard throughout the whole performance, which
lasted an hour and a half.   He did not once appear
to move, for he was so close to the skin behind
which Captain Lyon sat, that if he had done so he
must have perceived it.   Captain Lyon did not
hear the least rustling of his clothes, or even distinguish his breathing, although his outcries were
made with great exertion. †

* *Private Journal of* Captain G. F. Lyon.   Lond. 1824,
p. 358, 361.
† Id. Id. p. 366.

LETTER VIII.

Musical and harmonic sounds explained—Power of breaking glasses with the voice—Musical sounds from the vibration of a column of air—And of solid bodies—Kaleidophone—Singular acoustic figures produced on sand laid on vibrating plates of glass—and on stretched membranes—Vibration of flat rulers, and cylinders of glass—Production of silence from two sounds—Production of darkness from two lights—Explanation of these singular effects—Acoustic automaton—Droz's bleating sheep—Maillardet's singing bird—Vaucanson's flute player—His pipe and tabor player—Baron Kempelen's talking engine—Kratzenstein's speaking machine—Mr Willis's researches.

AMONG the discoveries of modern science there are few more remarkable than those which relate to the production of harmonic sounds.   We are all familiar with the effects of musical instruments, from the deep-toned voice of the organ to the wiry shrill of the Jew's harp.   We sit entranced under their magical influence, whether the ear is charmed with the melody of their sounds, or the heart agitated by the sympathies which they rouse.   But though we may admire their external form, and the skill of the artist who constructed them, we never think of inquiring into the cause of such extraordinary combinations.

Sounds of all kinds are conveyed to the organ of hearing through the air ; and if this element were to be destroyed, all nature would be buried in the deepest silence. Noises of every variety, whether they are musical or discordant, high or low, move through the air of our atmosphere at the surface of the earth with a velocity of 1090 feet in a second, or 765 miles per hour; but in sulphurous acid gas sound moves only through 751 feet in a second, while in hydrogen gas it moves with the great velocity of 3000 feet. Along fluid and solid bodies, its progress is still more rapid. Through water it moves at the rate of 4708 feet in a second, through tin at the rate of 8175 feet, and through iron, glass, and some kinds of wood, at the rate of 18530 feet.

When a number of single and separate sounds follow each other in rapid succession, they produce a continued sound, in the same manner as a continuous circle of light is produced by whirling round a burning stick before the eye. In order that the sound may appear a single one to the ear, nearly sixteen separate sounds must follow one another every second. When these sounds are exactly similar, and recur at equal intervals, they form a musical sound. In order to produce such sounds from the air, it must receive at least sixteen equally distant impulses or strokes in a second. The most common way of producing this effect is by a string or wire A B, Fig. 40, stretched between the fixed points A, B. If this string is taken by the middle and pulled aside, or if it is suddenly struck, it will vibrate between its two fixed points, as shown in the figure, passing alternately on each side of its axis A B, the vibrations gradually diminishing by the resistance of the air till the string is brought to rest. Its vibrations, however, may be kept up, by drawing a rosined fiddle-bow across it, and while it is vibrating it will give out a sound corresponding to the rapidity of its vibrations, and

Fig. 40.

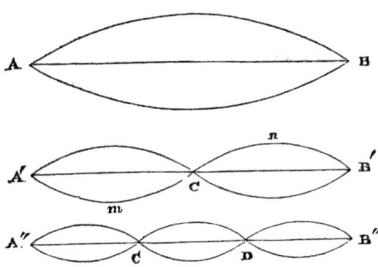

arising from the successive blows or impulses given to the air by the string.   This sound is called the fundamental sound of the string, and its acuteness or sharpness increases with the number of vibrations which the string performs in a second.

If we now touch the vibrating string A′ B′ lightly with the finger, or with a feather at the middle point C, Fig. 40, it will give out a more acute but fainter sound than before, and while the extent of its vibrations is diminished, their frequency is doubled.   In like manner, if we touch the string A″ B″, Fig 40, at a point C, so that A″C is one-third of A″ B″, the note will be still more acute, and correspond to thrice the number of vibrations.   All this might have been expected, but the wonderful part of the experiment is, that the vibrating string A′ B′ divides itself at C into two

parts A′C, CB′, the part A′C vibrating round A′ and
C as fixed points, and the part CB′ round C and B′,
but always so that the part A′ C is at the same
distance on the one side of the axis A′ B′ as at A
m C, while the part C B is on the other side, as at
C n B.   Hence the point C, being always pulled
by equal and opposite forces, remains at rest as if
it were absolutely fixed.   This stationary point is
called a *node,* and the vibrating portions A′ m C,
C n B′ loops.   The very same is true of the string
A″ B″, the points C and D being stationary points;
and upon the same principle a string may be divided into any number of vibrating portions.   In
order to prove that the string is actually vibrating
in these equal subdivisions, we have only to place
a piece of light paper with a notch in it on different
parts of the string.   At the nodes C and D it will
remain perfectly at rest, while at m or n in the
middle of the loops it will be thrown off or violently agitated.

The acute sounds given out by each of the vibrating portions are called *harmonic sounds,* and
they accompany the fundamental sound of the
string in the very same manner, as we have already seen, that the eye sees the accidental or
harmonic colours while it is affected with the fundamental colour.

The subdivision of the string, and consequently
the production of harmonic sounds, may be effected
without touching the string at all, and by means
of a sympathetic action conveyed by the air.   If a
string A B, for example, Fig. 40, is at rest, and if
a shorter string A″ C, one-third of its length, fixed at the two points A″ and C is set a-vibrating

in the same room, the string A B will be set a-vibrating in three loops like A″ B″, giving out the same harmonic sounds as the small string A″ C.

It is owing to this property of sounding bodies that singers with great power of voice are able to break into pieces a large tumbler glass, by singing close to it its proper fundamental note ; and it is from the same sympathetic communication of vibrations that two pendulum clocks fixed to the same wall, or two watches lying upon the same table, will take the same rate of going, though they would not agree with one another if placed in separate apartments. Mr Ellicott even observed that the pendulum of the one clock will stop that of the other, and that the stopped pendulum will after a certain time resume its vibrations, and in its turn stop the vibrations of the other pendulum.

The production of musical sounds by the vibrations of a column of air in a pipe is familiar to every person, but the extraordinary mechanism by which it is effected is known principally to philosophers. A column of air in a pipe may be set a-vibrating by blowing over the open end of it, as is done in Pan's pipes, or by blowing over a hole in its side as in the flute, or by blowing through an aperture called a reed, with a flexible tongue, as in the clarionet. In order to understand the nature of this vibration, let A B, Fig. 41, be a pipe or tube, and let us place in it a spiral spring A B, in which the coil or spire are at equal distances, each end of the spiral being fixed to the end of the tube. This elastic spring may be supposed to represent the air in the pipe, which is of equal density throughout. If we take hold of the spring at m, and push the point *m* towards A and towards B in succession, it will give us a good idea of the vibration of an elastic column of air.   When *m* is push-

Fig. 41.

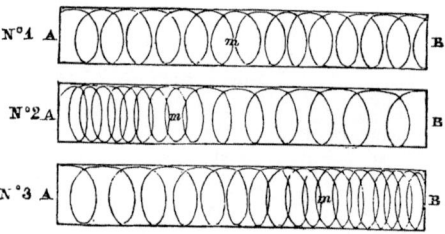

ed towards A, the spiral spring will be compressed or condensed, as shown at *m* A, No. 2, while at the other end, it will be dilated or rarefied, as shown at *m* B, and in the middle of the tube it will have the same degree of compression as in No. 1.   When the string is drawn to the other end of the tube B, the spring will be, as in No. 3, condensed at the end B, and dilated at the end A.   Now when a column of air vibrates in a pipe A B, the whole of it rushes alternately from B to A, as in No. 2, and from A to B as in No. 3, being condensed at the end A No. 2, and dilated or rarefied at the end B, while in No. 3 it is rarefied at A and condensed at B, preserving its natural density at the middle point between A and B.   In the case of the spring the ends A B are alternately pushed outwards and pulled inwards by the spring, the end A being pushed

outwards in No. 2, and B pulled inwards, while in
No. 3 A is pulled inwards and B pushed outwards.

That the air vibrating in a pipe is actually in
the state now described may be shown by boring
small holes in the pipe, and putting over them pieces
of a fine membrane. The membrane opposite to
the middle part between A and B, where the particles of the air have the greatest motion, will be
violently agitated, while at points nearer the ends
A and B it will be less and less affected.

Let us now suppose two pipes A B, B C, to be
joined together as in Fig. 42, and to be separated
by a fixed partition at B ; and let a spiral spring
be fixed in each. Let the spring AB be now
pushed to the end A, while the spring BC is
pushed to C, as in No. 1, and back again, as in No. 2,

Fig. 42.

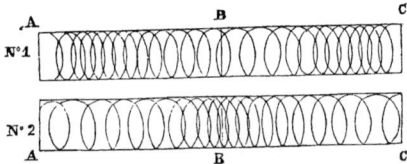

but always in opposite directions ; then it is obvious
that the partition B is in No. 1 drawn in opposite
directions towards A and towards C, and always
with forces equal to each other, that is, when B is
drawn slightly towards A, which it is at the beginning of the motion, it is also drawn slightly towards
C, and when it is drawn forcibly towards A, as it is
at the end of the motion of the spring, it is also
drawn forcibly towards C. If the partition B, there-

fore, is moveable, it will still remain fixed during
the opposite excursions of the spiral springs : Nay,
if we remove the partition, and hook the end of one
spiral spring to the end of the other, the node or
point of junction will remain stationary during the
movements of the springs, because at every instant
that point is drawn by equal and opposite forces.   If
three, four, or *five*, spiral springs are joined in a similar manner, we may conceive them all vibrating
between their nodes in the same manner.

Upon the very same principles we may conceive
a long column of air without partitions dividing itself into two, three, or four, smaller columns, each
of which will vibrate between its nodes in the same
manner as the spiral spring.   At the middle point
of each small vibrating column, the air will be of
its natural density like that of the atmosphere,
while at the nodes B, &c. it will be in a state of
condensation and rarefaction alternately.

If, when the air is vibrating in one column in the
pipe A B, as in Fig 41, No 2, 3, we conceive a hole
made in the middle, the atmospheric air will not
rush in to disturb the vibration, because the air
within the pipe and without it has exactly the same
density.   Nay, if, instead of a single hole, we were
to cut a ring out of the pipe at the middle point,
the column would vibrate as before.   But if we
bore a hole between the middle and one of the ends,
where the vibrating column must be either in a
state of condensation or rarefaction, the air must
either rush out or rush in, in order to establish the
equilibrium.   The air opposite the hole will then
be brought to the state of the external air like that
in the middle of the pipe, it will become the middle

of a vibrating column, and the whole column of air, instead of vibrating as one, will vibrate as two columns, each column vibrating with twice the velocity, and yielding harmonic sounds along with the fundamental sound of the whole columns, in the same manner as we have already explained with regard to vibrating strings. By opening other holes we may subdivide a vibrating column into any number of smaller vibrating columns. The holes in flutes, clarionets, &c. are made for this purpose. When they are all closed up the air vibrates in one column, and by opening and shutting the different holes in succession, the number of vibrating columns is increased or diminished at pleasure, and consequently the harmonic sounds will vary in a similar manner.

Curious as these phenomena are, they are still surpassed by those which are exhibited during the vibration of solid bodies. A rod or bar of metal or glass may be made to vibrate either longitudinally or laterally.

An iron rod will vibrate longitudinally like a column of air if we strike it at one end in the direction of its length, or rub it in the same direction with a wetted finger, and it will emit the same fundamental note as a column of air *ten* or *eleven* times as long, because sound moves as much faster in iron than in air. When the iron rod is thus vibrating along its length, the very same changes which we have shown in Fig. 41, as produced in a spiral spring or in a column of air, take place in the solid metal. All its particles move alternately towards A and towards B, the metal being in the one case condensed at the end to which the particles move, and expanded

at the end from which they move, and retaining its
natural density in the middle of the rod. If we now
hold this rod in the middle, by the finger and thumb
lightly applied, and rub it in the middle either of
A B or B C with a piece of cloth sprinkled with
powdered rosin, or with a well-rosined fiddle-bow
drawn across the rod, it will divide itself into two
vibrating portions A B, B C, each of which will vibrate, as shown in Fig. 42, like the two adjacent columns of air, the section of the rod, or the particles
which compose that section at B, being at perfect
rest. By holding the rod at any intermediate point
between A and B, so that the distance from A to
the finger and thumb is one-third, one-fourth, one-
fifth, &c. of the whole length A C, and rubbing one
of the divisions in the middle, the rod will divide itself into 3, 4, 5, &c. vibrating portions, and give out
corresponding harmonic sounds.

A rod of iron may be made to vibrate laterally
or transversely by fixing one end of it firmly as in
a vice, and leaving the other free, or by having both
ends free or both fixed. When a rod, fixed at one
end and free at the other, is made to vibrate, its
mode of vibrating may be rendered evident to the
eye ; and for the purpose of doing this Mr Wheat-
stone has contrived a curious instrument called the
Kaleidophone, which is shown in Fig. 43. It consists of a circular base of wood A B, about *nine* inches in diameter and one inch thick, and having four
brass sockets firmly fixed into it at C, D, E, and F,
Into these sockets are screwed four vertical steel
rods C, D, E, and F, about 13 or 14 inches long,
one being a square rod, another a bent cylindrical one,

and the other two cylindrical ones of different diameters. On the extremities of these rods are fix-

Fig. 43.

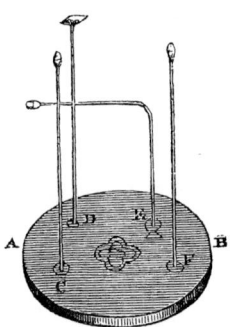

ed small quicksilvered glass beads, either singly or
in groups, so that when the instrument is placed
in the light of the sun or in that of a lamp, bright
images of the sun or candle are seen reflected on
each bead. If any of these rods is set a-vibrating,
these luminous images will form continuous and
returning curve lines in a state of constant variation, each different rod giving curves of different characters.

The melodion, an instrument of great power,
embracing five octaves, operates by means of the
vibrations of metallic rods of unequal lengths, fixed at one end, and free at the other. * A narrow
and thin plate of copper is screwed to the free extremity of each rod, and at right angles to its length;

* See Edinburgh Encyclopædia, Art. Science, Curiosities
in, Vol. xvii. p. 563.

and its surface is covered with a small piece of felt impregnated with rosin. This narrow band is placed near the circumference of a revolving cylinder, and by touching the key it is made to descend till it touches the revolving cylinder, and gives out its sound. The sweetness and power of this instrument are unrivalled, and such is the character of its tones, that persons of a nervous temperament are often entirely overpowered by its effects.

The vibrations of plates of metal or glass of various forms exhibit a series of the most extraordinary phenomena which are capable of being shown by very simple means. These phenomena are displayed in an infinite variety of regular figures assumed by sand, or fine lycopodium powder, strewed over the surface of the glass plate. In order to produce these figures, we must pinch or damp the plate at one or more places, and when the sand is strewed upon its surface, it is thrown into vibrations by drawing a fiddle-bow over different parts of its circumference. The method of damping or pinching plates is shown in Fig. 44. In No. 1, a square plate of glass A B, ground smooth at its edges, is pinched by the finger and thumb. In No. 2 a circular plate is held by the thumb against the top c of a perpendicular rod, and damped by the fingers at two different points of its circumference. In No. 3 it is damped at three points of its circumference, c and d by the thumb and finger, and at e by pressing it against a fixed obstacle a b. By a means of a clamp like that at No. 4, it may be damped at a greater number of points.

If we take a *square* plate of glass, such as that shown in Fig. 45, No. 1, and pinching it at its

4

centre, draw the fiddle-bow near one of its angles, the sand will accumulate in the form of a cross, as

Fig. 44.

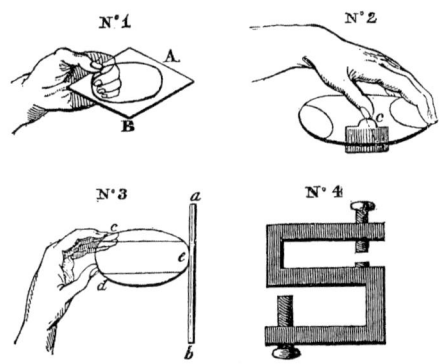

shown in the figure, being thrown off the parts of the plate that are in a state of vibration, and settling in the nodes or parts which are at rest. If the bow is drawn across the middle of one of the edges, the sand will accumulate as in No. 2. If the plate is pinched at N, No. 3, and the bow applied at F and perpendicular to A B, the sand will arrange itself in three parallel lines, perpendicular to a fourth passing through F and N. But if the point N where it is pinched, is a little farther from the edge than in No. 3, the parallel lines will change into curves as in No. 4.

If the plate of glass is circular and pinched at its centre, and also at a point of its circumference, and if the bow is applied at a point 45° from the last

point, the figure of the sand will be as in Fig. 46,
No. 1. If with the same plate, similarly pinched,

Fig. 45.

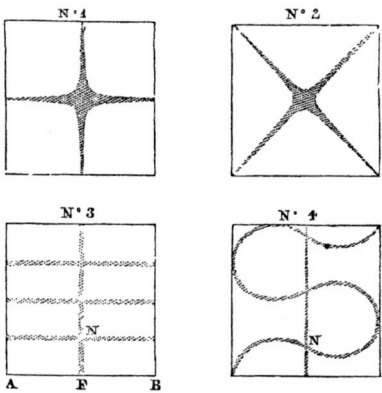

the bow is drawn over a part 30° from the pinched
point of the circumference, the sand will form six
radii as in No. 2. When the centre of the plate is
left free, a different set of figures is produced as
shown in No. 3 and No. 4. When the plate is
pinched near its edge, and the bow applied 45° from
the point pinched, a circle of sand will pass through
that point, and two diameters of sand at right
angles to each other, will be formed, as in No. 3.
When a point of the circumference is pressed against
a fixed obstacle, and the bow applied 30° from that
point, the figure in No. 4 is produced.

3

If, in place of a solid plate, we strew the sand over a stretched membrane, the sand will form it-

Fig. 46.

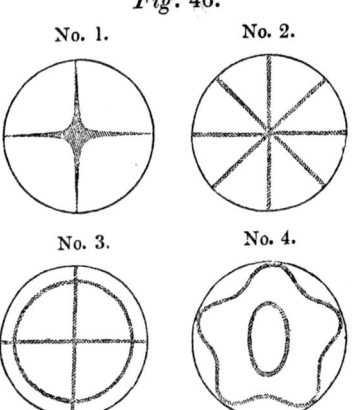

If, in place of a solid plate, we strew the sand

No. 1. No. 2.

No. 3. No. 4.

self into figures, even when the vibrations are communicated to the membrane through the air. In order to make these experiments, we must stretch a thin sheet of wet paper, such as vegetable paper, over the mouth of a tumbler-glass with a footstalk, and fix it to the edges with glue. When the paper is dry, a thin layer of dry sand is strewed upon its surface. If we place this membrane upon a table, and hold immediately above it and parallel to the membrane, a plate of glass vibrating so as to give any of the figures shown in Fig. 46, the sand upon the membrane will imitate exactly the figure upon the glass. If the glass plate, in place of vibrating horizontally, is made to vibrate in an

N

inclined position, the figures on the membrane will change with the inclination, and the sand will assume the most curious arrangements. The figures thus produced vary with the size of the membrane, with its material, its tension, and its shape. When the same figure occurred several times in succession, a breath upon the paper will change its degree of tension, and produce an entirely new figure, which, as the temporary moisture evaporates, will return to the original figure, through a number of intermediate ones. The pipe of an organ at the distance of a few feet, or the notes of a flute at the distance of half a foot, will arrange the sand on the membrane into figures which perpetually change with the sound that is produced.

The manner in which flat rulers and cylinders of glass perform their vibrations is very remarkable. If a glass plate about twenty-seven inches long, six-tenths of an inch broad, and six-hundredths of an inch thick, is held by the edges between the finger and thumb, and has its lower surface, near either end, rubbed with a piece of wet cloth, sand laid upon its upper surface will arrange itself in parallel lines at right angles to the length of the plate. If the place of these lines is marked with a dot of ink, and the other side of the glass ruler is turned upwards, and the ruler made to vibrate as before, the sand will now accumulate in lines intermediate between the former lines, so that the motions of one-half the thickness of the glass ruler are precisely the reverse of those of the corresponding parts of the other half.

As these singular phenomena have not yet been made available by the scientific conjuror, we must

be satisfied with this brief notice of them ; but
there is still one property of sound, which has its
analogy also in light, too remarkable to be passed
without notice.   This property has more of the
marvellous in it than any result within the wide
range of the sciences.   *Two loud sounds may be
made to produce silence,* and *two strong lights may
be made to produce darkness !*

If two equal and similar strings, or the columns
of air in two equal and similar pipes, perform exactly 100 vibrations in a second, they will produce
each equal waves of sound, and these waves will
conspire in generating an uninterrupted sound,
double of either of the sounds heard separately.   If
the two strings or the two columns of air are not
in unison, but nearly so, as in the case where the
one vibrates 100, and the other 101 times in a second, then at the first vibration the two sounds will
form one of double the strength of either ; but the
one will gradually gain upon the other, till at the
fiftieth vibration it has gained half a vibration on
the other.   At this instant the two sounds will *destroy one another,* and an interval of perfect silence
will take place.   The sound will instantly commence, and gradually increase till it becomes loudest
at the hundredth vibration, where the two vibrations conspire in producing a sound double of
either.   An interval of silence will again occur at
the 150th, 250th, 350th vibration, or every second,
while a sound of double the strength of either will be
heard at the 200dth, 300dth, and 400dth vibration.
When the unison is very defective, or when there
is a great difference between the number of vibrations which the two strings or columns of air per-

form in a second, the successive sounds and intervals of silence resemble a rattle. With a powerful organ the effect of this experiment is very
fine, the repetition of the sounds *wow—wow—
wow*,—representing the doubled sound and the interval of silence which arise from the total extinction of the two separate sounds.

The phenomena corresponding to this in the case
of light, is perhaps still more surprising. If a
beam of *red* light issues from a luminous point, and
falls upon the retina, we shall see distinctly the luminous object from which it proceeds ; but if another
pencil of red light issues from another luminous
point any how situated, provided the difference between its distance and that of the other luminous
point from the point of the retina, on which the first
beam fell, is the 258th thousandth part of an inch,
or exactly *twice, thrice, four* times &c. that distance ;
and if this second beam falls upon the same point
of the retina, the one light will increase the intensity of the other, and the eye will see *twice* as much
light as when it received only one of the beams
separately. All this is nothing more than what
might be expected from our ordinary experience.
But if the difference in the distances of the two
luminous points is only *one-half* of the 258th thou-
sandth part of an inch, or $1\frac{1}{2}$, $2\frac{1}{2}$, $3\frac{1}{2}$, $4\frac{1}{2}$, times
that distance, *the one light will extinguish the other,
and produce absolute darkness*. If the two luminous points are so situated, that the difference of
their distances from the point of the retina, is intermediate between 1 and $1\frac{1}{2}$, or 2 and $2\frac{1}{2}$, above
the 258th thousandth part of an inch, the intensity
of the effect which they produce will vary from

absolute darkness to double the intensity of either light. At $1\frac{1}{4}$, $2\frac{1}{4}$, $3\frac{1}{4}$, times, &c. the 258th thousandth of an inch, the intensity of the two combined lights will be equal only to one of them acting singly. If the lights, in place of falling upon the retina, fall upon a sheet of white paper, the very same effect will be produced, a black spot being produced in the one case, and a bright white one in the other, and intermediate degrees of brightness in intermediate cases. If the two lights are *violet*, the difference of distances at which the preceding phenomena will be produced, will be the 157th thousandth part of an inch, and it will be intermediate between the 258th and the 157th thousand part of an inch, for the intermediate colours. This curious phenomenon may be easily shown to the eye, by admitting the sun's light into a dark room through a small hole about the 40th or 50th part of an inch in diameter, and receiving the light on a sheet of paper. If we hold a needle or piece of slender wire in this light, and examine its shadow, we shall find that the shadow consists of bright and dark stripes succeeding each other alternately, the stripe in the very middle or axis of the shadow being a bright one. The rays of light which are bent into the shadow, and which meet in the very middle of the shadow, have exactly the same length of path, so that they form a bright fringe of double the intensity of either; but the rays which fall upon a point of the shadow at a certain distance from the middle, have a difference in the length of their paths, corresponding to the difference at which the lights destroy each other, so that a *black* stripe is produced on each side of the middle bright one. At a greater distance from the

middle, the difference becomes such as to produce
a bright stripe, and so on, a bright and a dark stripe
succeeding each other to the margin of the shadow.

The explanation which philosophers have given
of these strange phenomena is very satisfactory, and
may be easily understood. When a wave is made
on the surface of a still pool of water, by plunging
a stone into it, the wave advances along the surface, while the water itself is never carried forward,
but merely rises into a height and falls into a hollow, each portion of the surface experiencing an
elevation and a depression in its turn. If we suppose two waves equal and similar to be produced
by two separate stones, and if they reach the same
spot at the same time, that is, if the two elevations
should exactly coincide, they would unite their
effects and produce a wave twice the size of either;
but if the one wave should be just so far before the
other, that the hollow of the one coincided with
the elevation of the other, and the elevation of the
one with the hollow of the other, the two waves
would obliterate or destroy one another, the elevation as it were of the one filling up half the hollow
of the other, and the hollow of the one taking away
half the elevation of the other, so as to reduce the
surface to a level. These effects will be actually exhibited by throwing two equal stones into a pool of
water, and it will be seen that there are certain lines
of a hyperbolic form where the water is quite
smooth, in consequence of the equal waves oblite-
rating one another, while in other adjacent parts,
the water is raised to a height corresponding to both
the waves united.

In the tides of the ocean, we have a fine example

of the same principle. The two immense waves arising from the action of the sun and moon upon the ocean, produce our spring-tides by their combination, or when the elevations of each coincide, and our neap-tides, when the elevation of the one wave coincides with the depression of the other. If the sun and moon had exerted exactly the same force upon the ocean, or produced tide waves of the same size, then our neap-tides would have disappeared altogether, and the spring-tide would have been a wave double of the wave produced by the sun and moon separately. An example of the effect of the equality of the two waves occurs in the port of Batsha, where the two waves arrive by channels of different lengths, and actually obliterate each other.

Now, as sound is produced by undulations or waves in the air, and as light is supposed to be produced by waves or undulations in an etherial medium, filling all nature, and occupying the pores of transparent bodies, the successive production of sound and silence, by two loud sounds, or of light and darkness by two bright lights, may be explained in the very same manner as we have explained the increase and the obliteration of waves formed on the surface of water. If this theory of light be correct, then the breadth of a wave of *red* light will be the 258th thousandth part of an inch, the breadth of a wave of green light the 207th thousandth part of an inch, and the breadth of a wave of violet light, the 157th thousandth part of an inch.

Among the wonders of modern skill, we must enumerate those beautiful automata by which the motions and actions of man and other animals have been successfully imitated. I shall therefore describe at present some of the most remarkable acoustic automata, in which the production of musical and vocal sounds has been the principal object of the artist.

Many very ingenious pieces of acoustic mechanism have been from time to time exhibited in Europe. The celebrated Swiss mechanist, M. Le Droz, constructed for the King of Spain the figure of a sheep, which imitated in the most perfect manner the bleating of that animal, and likewise the figure of a dog watching a basket of fruit, which, when any of the fruit was taken away, never ceased barking till it was replaced.

The singing-bird of M. Maillardet, which he exhibited in Edinburgh many years ago, is still more wonderful.* An oval box, about three inches long, was set upon the table, and in an instant the lid flew up, and a bird of the size of the humming-bird, and of the most beautiful plumage, started from its nest. After fluttering its wings, it opened its bill and performed four different kinds of the most beautiful warbling. It then darted down into its nest, and the lid closed upon it. The moving power in this piece of mechanism is said to have been springs which continued their action only four minutes. As there was no room, within so small a figure, for accommodating pipes to produce the great variety of notes which were warbled, the artist used only one tube, and produced all the variety of sounds by shortening and lengthening it with a moveable piston.

Ingenious as these pieces of mechanism are, they sink into insignificance when compared with the

* A similar piece of mechanism had been previously made by M. le Droz.

machinery of M. Vaucanson, which had previously
astonished all Europe. His two principal automata were the flute-player, and the pipe and tabor-
player. The flute-player was completed in 1736,
and wherever it was exhibited it produced the
greatest sensation. When it came to Paris it was
received with great suspicion. The French sca-
vans recollected the story of M. Raisin, the organist
of Troyes, who exhibited an automaton player upon
the harpsichord, which astonished the French court
by the variety of its powers. The curiosity of the
King could not be restrained, and in consequence
of his insisting upon examining the mechanism,
there was found in the figure a pretty little musi-
cian five years of age. It was natural, therefore,
that a similar piece of mechanism should be received
with some distrust ; but this feeling was soon removed by M. Vaucanson, who exhibited and explained to a committee of the Academy of Sciences the
whole of the mechanism. This learned body was
astonished at the ingenuity which it displayed; and
they did not hesitate to state, that the machinery
employed for producing the sounds of the flute,
performed in the most exact manner the very operations of the most expert flute-player, and that the
artist had imitated the effects produced, and the
means employed by nature, with an accuracy which
exceeded all expectation. In 1738, M. Vaucanson
published a memoir, approved of by the academy,
in which he gave a full description of the machinery
employed, and of the principles of its construction.
Following this memoir, I shall therefore attempt to
give as popular a description of the automaton as can

be done without lengthened details and numerous figures.

The body of the flute-player was about $5\frac{1}{2}$ feet high, and was placed upon a piece of rock, surrounding a square pedestal $4\frac{1}{2}$ feet high by three and a half wide. When the panel which formed the front of the pedestal was opened, there was seen on the right a clock movement, which, by the aid of several wheels, gave a rotatory motion to a steel axis about $2\frac{1}{2}$ feet long, having cranks at six equidistant points of its length, but lying in different directions. To each crank was attached a cord, which descended and was fixed by its other end to the upper board of a pair of bellows, $2\frac{1}{2}$ feet long and six inches wide. Six pair of bellows arranged along the bottom of the pedestal were then wrought, or made to blow in succession, by turning the steel axis.

At the upper face of the pedestal, and upon each pair of bellows, is a double pulley, one of whose rims is 3 inches in diameter, and the other $1\frac{1}{2}$. The cord which proceeds from the crank coils round the smallest of these pulleys, and that which is fixed to the upper board of the bellows goes round the larger pulley. By this means the upper board of the bellows is made to rise higher than if the cords went directly from them to the cranks.

Round the larger rims of three of these pulleys, viz. those on the right hand, there are coiled three cords, which, by means of several smaller pulleys, terminate in the upper boards of other three pair of bellows placed on the top of the box.

The tension of each cord when it begins to raise the board of the bellows to which it is attached,

gives motion to a lever placed above it between the
axis and the double pulley in the middle and
lower region of the box.   The other end of this
lever keeps open the valve in the lower board of
the bellows, and allows the air to enter freely, while
the upper board is rising to increase the capacity
of the bellows.   By this means there is not only
power gained, in so far as the air gains easier admission through the valve, but the fluttering noise
produced by the action of the air upon the valves
is entirely avoided, and the nine pair of bellows are
wrought with great ease, and without any concus-
sion or noise.

These nine bellows discharge their wind into
three different and separate tubes.   Each tube receives the wind of three bellows, the upper boards
of one of the three pair being loaded with a weight
of four pounds, those of the second three pair with
a weight of two pounds, and those of the other
three pair with no weight at all.   These three tubes
ascended through the body of the figure, and terminated in three small reservoirs placed in its
trunk.   These reservoirs were thus united into one,
which, ascending into the throat, formed by its en-
largement the cavity of the mouth terminated by
two small lips, which rested upon the hole of the
flute.   These lips had the power of opening more
or less, and by a particular mechanism, they could
advance or recede from the hole in the flute.   Within
the cavity of the mouth there is a small moveable
tongue for opening and shutting the passage for the
wind through the lips of the figure.

The motions of the fingers, lips, and tongue, of
the figure, were produced by means of a revolving

cylinder thirty inches long, and twenty-one in diameter.   By means of pegs and brass staples fixed
in fifteen different divisions in its circumference,
fifteen different levers, similar to those in a barrel-
organ, were raised and depressed.   Seven of these
regulated the motions of the seven fingers for stop-
ping the holes of the flute, which they did by means
of steel chains rising through the body and directed by pulleys to the shoulder, elbow, and fingers.
Other three of the levers communicating with the
valves of the three reservoirs, regulated the ingress
of the air, so as to produce a stronger or a weaker
tone.   Another lever opened the lips so as to give
a free passage to the air, and another contracted
them for the opposite purpose.   A third lever drew
them backwards from the orifice of the flute, and a
fourth pushed them forward.   The remaining lever
enabled the tongue to stop up the orifice of the
flute.

Such is a very brief view of the general mechanism
by which the requisite motions of the flute-player
were produced.   The airs which it played were
probably equal to those executed by a living performer, and its construction, as well as its performances, continued for many years to delight and
astonish the philosophers and musicians of Europe.

Encouraged by the success of this machine, M.
Vaucanson exhibited in 1741 other automata, which
were equally, if not more, admired.   One of these
was the automaton duck, which performed all
the motions of that animal, and not only ate its
food, but digested it ; * and the other was his pipe

* See Letter XI.

and tabor-player, a piece of mechanism which required all the resources of his fertile genius. Having begun this machine before he was aware of its peculiar difficulties, he was often about to abandon it in despair, but his patience and his ingenuity combined, enabled him not only to surmount every difficulty, but to construct an automaton which performed complete airs, and greatly excelled the most esteemed peformers on the pipe and tabor.

The figure stands on a pedestal, and is dressed like a dancing shepherd. He holds in one hand a flageolet, and in the other the stick with which he beats the tambourin as an accompaniment to the airs of the flageolet, about twenty of which it is capable of performing. The flageolet has only three holes, and the variety of its tones depends principally on a proper variation of the force of the wind, and on the different degrees with which the orifices are covered. These variations in the force of the wind required to be given with a rapidity which the ear can scarcely follow, and the articulation of the tongue was required for the quickest notes, otherwise the effect was far from agreeable. As the human tongue is not capable of giving the requisite articulations to a rapid succession of notes, and generally slurs over one-half of them, the automaton was thus able to excel the best performers, as it played complete airs with articulations of the tongue at every note.

In constructing this machine M. Vaucanson observed that the flageolet must be a most fatiguing instrument for the human lungs, as the muscles of the chest must make an effort equal to 56 pounds in order to produce the highest notes. A single ounce was sufficient for the lowest notes, so that we may, from this circumstance, form an idea of the variety of intermediate effects required to be produced.

While M. Vaucanson was engaged in the construction of these wonderful machines, his mind was filled with the strange idea of constructing an automaton containing the whole mechanism of the circulation of the blood. From some birds which he made he was satisfied of its practicability ; but as the whole vascular system required to be made of elastic gum or caoutchouc, it was supposed that it could only be executed in the country where the caoutchouc tree was indigenous. Louis XVI. took a deep interest in the execution of this machine. It was agreed that a skilful anatomist should proceed to Guyana to superintend the construction of the blood-vessels, and the King had not only approved of, but had given orders for, the voyage. Difficulties, however, were thrown in the way : Vaucanson became disgusted, and the scheme was abandoned.

The two automata which we have described were purchased by Professor Bayreuss of Helmstadt; but we have not been able to learn whether or not they still exist.

Towards the end of the seventeenth century a bold and almost successful attempt was made to construct a *talking automaton*. In the year 1779, the Imperial Academy of Sciences at St Petersburgh proposed as the subject of one of their annual prizes an inquiry into the nature of the vowel sounds, A, E, I, O and U, and the construction of an instrument for artificially imitating them. This prize was

3

gained by M. Kratzenstein, who showed that all
the vowels could be distinctly pronounced by blowing through a reed into the lower ends of the pipes
of the annexed figures, as shown in Fig. 47, where
the corresponding vowels are marked on the diffe-

Fig. 47.

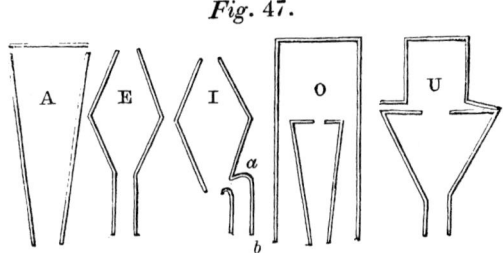

rent pipes. The vowel I is pronounced by merely
blowing into the pipe *a b*, of the pipe marked I,
without the use of a reed.

About the same time that Kratzenstein was engaged in these researches, M. Kempelen of Vienna, a
celebrated mechanician, was occupied with the same
subject. In his first attempt he produced the vowel
sounds, by adapting a reed R, Fig. 48, to the bottom of a funnel-shaped cavity A B, and placing his
hand in various positions within the funnel. This
contrivance, however, was not fitted for his purpose,
but after long study, and a diligent examination of
the organs of speech, he contrived a hollow oval
box, divided into two portions attached by a hinge
as so to resemble jaws. This box received the
sound which issued from the tube connected with
the reed, and by opening and closing the jaws, he
produced the sounds, A, O, O U, and an imperfect

E, but no indications of an I.   After two years labour he succeeded in obtaining from different jaws

Fig. 48.

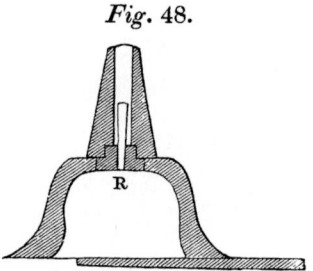

the sounds of the consonants P, M, L, and by
means of these vowels, and consonants, he could
compose syllables and words, such as *mama, papa,
aula, lama, mulo*.   The sounds of two adjacent
letters, however, run into each other, and an aspira-
tion followed some of the consonants, so that instead of *papa* the word sounded *phaa-ph-a ;* these
difficulties he contrived with much labour to surmount, and he found it necessary to imitate the
human organs of speech by having only one mouth
and one glottis.   The mouth consisted of a funnel
or bell-shaped piece of elastic gum, which approxi-
mated, by its physical properties, to the softness
and flexibility of the human organs.*   To the
mouth-piece was added a nose made of two tin

* Had M. Kempelen known the modern discovery of
giving caoutchouc any degree of softness, by mixing it with
molasses or sugar, which is always absorbing moisture from
the atmosphere, he might have obtained a still more perfect
imitation of the human organs.

tubes, which communicated with the mouth. When both these tubes were open, and the mouth-piece closed, a perfect M was produced, and when one was closed and the other open, an N was sounded. M. Kempelen could have succeeded in obtaining the four letters D, G, K, T, but by using a P instead of them, and modifying the sound in a particular manner, he contrived to deceive the ear by a tolerable resemblance of these letters.

There seems to be no doubt that he at last was able to produce entire words, and sentences such as, *opera, astronomy, Constantinopolis, vous etés mon ami, je vous aime de tout mon cœur, venez avec moi à Paris, Leopoldus secundus, Romanorum imperator semper Augustus, &c.* but he never fitted up a speaking figure, and probably, from being dissatisfied with the general result of his labours, he exhibited only to his private friends the effects of the apparatus, which was fitted up in the form of a box.

This box was rectangular, and about three feet long, and was placed upon a table and covered with a cloth. When any particular word was mentioned by the company, M. Kempelen caused the machine to pronounce it, by introducing his hands beneath the cloth, and apparently giving motion to some parts of the apparatus. Mr Thomas Collinson, who had seen this machine in London, mentions in a letter to Dr Hutton, that he afterwards saw it at M. Kempelen's own house in Vienna, and that he then gave it the same word to be pronounced, which he gave it in London, viz. the word *Exploitation*, which, he assures us, it again distinctly pronounced with the French accent.

o

M. Kratzenstein seems to have been equally unsuccessful, for though he assured M. De Lalande,
when he saw him in Paris in 1786, that he had
made a machine which could speak pretty well, and
though he showed him some of the apparatus by
which it could sound the vowels, and even such
syllables as *papa* and *mama*, yet there is no reason
to believe that he had accomplished more than this.

The labours of Kratzenstein and Kempelen
have been recently pursued with great success by
our ingenious countryman Mr Willis of Cambridge.
In repeating Kempelen's experiment shown in Fig.
48, he used a shallower cavity, such as that in Fig.
49, and found that he could entirely dispense with
the introduction of the hand, and could obtain the

Fig. 49.

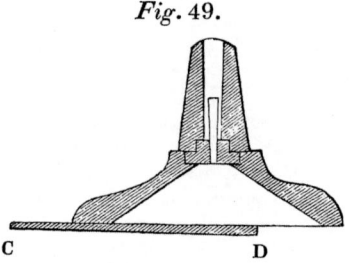

C   D

whole series of vowels, by sliding a flat board C D
over the mouth of the cavity. Mr· Willis then
conceived the idea of adapting to the reed cylindrical tubes, whose length could be varied by sliding
joints. When the tube was greatly less than the
length of a stopped pipe in unison with the reed, it
sounded I, and by increasing the length of the tube
it gave E, A, O and U in succession. But what

was very unexpected, when the tube was so much lengthened as to be $1\frac{1}{2}$ times the length of a stopped pipe in unison with the reed, the vowels began to be again sounded in an inverted order, viz. U, O, A, E, and then again in a direct order, I, E, A, O, U, when the length of the tube was equal to twice that of a stopped pipe, in unison with the reed.

Some important discoveries have been recently made by M. Savart respecting the mechanism of the human voice,* and we have no doubt that, before another century is completed, a *Talking* and a *Singing machine* will be numbered among the conquests of Science.

* See Edinburgh Journal of Science, No. viii. p. 200.

LETTER IX.

ALTHOUGH among the phenomena of the material world there is scarcely one, which, when well considered, is not an object of wonder, yet those which we have been accustomed to witness from our infancy lose all their interest from the frequency of their occurrence, while to the natives of other countries they are unceasing objects of astonishment and delight. The inhabitant of a tropical climate is confounded at the sight of falling snow, and he almost discredits the evidence of his senses, when he sees a frozen river carrying loaded waggons on its surface. The diffusion of knowledge by books, as well as by frequent communication between the natives of different quarters of the globe, have deprived this class of local wonders of their influence, and the Indian and the Scandinavian can visit each other's lands without any violent excitement of surprise. Still, however, there are phenomena of rare occurrence, of which no description can convey the idea, and which continue to be as deeply marked with the marvellous as if they had been previously unknown. Among these we may rank the remarkable modifications which sound undergoes in particular situations and under particular circumstances.

In the ordinary intercourse of life, we recognize individuals as much by their voice as by the features of their face and the form of their body. A friend who has been long absent will often stand before us as a stranger, till his voice supplies us with the full power of recognition. The brand imprinted by time on his outer form may have effaced the youthful image which the memory had cherished, but the original character of his voice and its yet remembered tones will remain unimpaired.

An old friend with a new face is not more common in its moral than in its physical acceptation, and though the sagacity of proverbial wisdom has not supplied us with the counterpart in relation to the human voice, yet the influence of its immutability over the mind has been recorded by the poet in some of his most powerful conceptions. When Manfred was unable to recognize in the hectic phantom of Astarte the endeared lineaments of the being whom he loved, the mere utterance of his name recalled " the voice which was his music," and invested her with the desired reality.

Say on, say on—
I live but in the Sound—It is thy voice!

BYRON.

The permanence of character thus impressed upon speech exists only in those regions to whose atmosphere our vocal organs are adapted. If either the speaker or the hearer is placed in air differing greatly in density from that to which they are accustomed, the voice of the one will emit different sounds, or the same sounds will produce a different impression on the ear of the other. But if both parties are placed in this new atmosphere, their tones of communication will suffer the most remarkable change. The two extreme positions, where such effects become sufficiently striking, are in the compressed air of the diving-bell, when it is immersed to a great depth in the sea, or in the rarefied atmosphere which prevails on the summit of the Himalaya or the Andes.

In the region of common life, and even at the stillest hour of night, the ear seldom rests from its toils. When the voice of man and the bustle of his labours have ceased, the sounds of insect life are redoubled, the night breeze awakens among the rustling leaves, and the swell of the distant ocean, and the sounds of the falling cataract or of the murmuring brook, fill the air with their pure and solemn music. The sublimity of deep silence is not to be found even in the steppes of the Volga, or in the forests of the Orinoco. It can be felt only in those lofty regions

Where the tops of the Andes
Shoot soaringly forth.

As the traveller rises above the limit of life and motion, and enters the region of habitual solitude, the death-like silence which prevails around him is rendered still more striking by the diminished density of the air which he breathes. The voice of his fellow traveller ceases to be heard even at a moderate distance, and sounds which would stun the ear at a lower level make but a feeble impression. The report of a pistol on the top of Mont Blanc is no louder than that of an Indian cracker. But while the thinness of the air thus subdues the loudest sounds, the voice itself undergoes a singular change: the muscular energy by which we speak experiences a great diminution, and our powers of utterance, as well as our power of hearing, are thus singularly modified. Were the magician, therefore, who is desirous to impress upon his victim or upon his pupil the conviction of his supernatural power, to carry him, under the injunction of silence,

————————————————to breathe
The difficult air of the iced mountain's top,
Where the birds dare not build, nor insects wing
Flit o'er the herbless granite,

he would experience little difficulty in arresting his power over the elements, and still less in subsequently communicating the same influence to his companion.

But though the air at the tops of our highest mountains is scarcely capable of transmitting sounds of ordinary intensity, yet sounds of extraordinary power force their way through its most attenuated strata. At elevations where the air is three thousand times more rare than that which we breathe, the explosion of meteors is heard like the sound of cannon on the surface of the earth, and the whole air is often violently agitated by the sound. This fact alone may give us some idea of the tremendous nature of the forces which such explosions create, and it is fortunate for our species, that they are confined to the upper regions of the atmosphere. If the same explosions were to take place in the dense air which rests upon the earth, our habitations and our lives would be exposed to the most imminent peril.

Buildings have often been thrown down by violent concussions of the air, occasioned either by the sound of great guns, or by loud thunder, and the most serious effects upon human and animal life have been produced by the same cause. Most persons have experienced the stunning pain produced in the ear, when placed near a cannon that is discharged. Deafness has frequently been the result of such sudden concussions, and if we may reason from analogy, death itself must often have been the consequence. When peace was proclaimed in London in 1697, two troops of horse were dismounted and drawn up in line in order to fire their vollies. Opposite the centre of the line was the door of a butcher's shop, where there was a large mastiff dog of great courage. This dog was sleeping by the fire, but when the first volley was fired, it immediately started up, ran into another room, and hid itself under a bed. On the firing of the second volley, the dog rose, run several times about the room trembling violently and apparently in great agony. When the third volley was fired, the dog

ran about once or twice with great violence, and
instantly fell down dead, throwing up blood from
his mouth and nose.

Sounds of known character and intensity are
often singularly changed even at the surface of the
earth according to the state of the ground and the
conditions of the clouds.   On the extended heath,
where there are no solid objects capable of reflecting
or modifying sound, the sportsman must frequently
have noticed the unaccountable variety of sounds
which are produced by the report of his fowling-piece.
Sometimes they are flat and prolonged, at other times
short and sharp, and sometimes the noise is so strange
that it is referred to some mistake in the loading of
the gun.   These variations, however, arise entirely
from the state of the air, and from the nature and
proximity of the superjacent clouds.   In pure air
of uniform density the sound is sharp and soon over,
as the undulations of the air advance without any
interrupting obstacles.   In a foggy atmosphere, or
where the vapours produced by heat are seen dancing as it were in the air, the sound is dull and prolonged, and when these clouds are immediately over
head, a succession of echoes from them produces a
continued or a reverberating sound.   When the
French astronomers were determining the velocity
of sound by firing great guns, they observed that the
report was always single and sharp under a perfectly
clear sky, but indistinct, and attended by a long
continued roll like thunder, when a cloud covered a
considerable part of the horizon.   It is no doubt
owing to the same cause, namely, the reflexion
from the clouds, that the thunder rolls through the

heavens, as if it were produced by a succession of electric explosions.

The great audibility of sounds during the night is a phenomenon of considerable interest, and one which had been observed even by the ancients. In crowded cities or in their vicinity, the effect was generally ascribed to the rest of animated beings, while in localities where such an explanation was inapplicable, it was supposed to arise from a favourable direction of the prevailing wind. Baron Humboldt was particularly struck with this phenomenon when he first heard the rushing of the great cataracts of the Orinoco in the plain which surrounds the mission of the Apures. These sounds he regarded as three times louder during the night than during the day. Some authors ascribed this fact to the cessation of the humming of insects, the singing of birds, and the action of the wind on the leaves of the trees, but M. Humboldt justly maintains that this cannot be the cause of it on the Orinoco, where the buzz of insects is much louder in the night than in the day, and where the breeze never rises till after sunset. Hence he was led to ascribe the phenomenon to the perfect transparency and uniform density of the air, which can exist only at night after the heat of the ground has been uniformly diffused through the atmosphere. When the rays of the sun have been beating on the ground during the day, currents of hot air of different temperatures, and consequently of different densities, are constantly ascending from the ground and mixing with the cold air above. The air thus ceases to be a homogeneous medium, and every person must have observed the effects of it upon objects seen through it which

are very indistinctly visible, and have a tremulous
motion, as if they were " dancing in the air." The
very same effect is perceived when we look at objects
through spirits and water that are not perfectly
mixed, or when we view distant objects over a red
hot poker or over a flame. In all these cases the
light suffers refraction in passing from a medium
of one density into a medium of a different density,
and the refracted rays are constantly changing their
direction as the different currents rise in succession. Analogous effects are produced when sound
passes through a mixed medium, whether it consists of two different mediums or of one medium
where portions of it have different densities. As
sound moves with different velocities through media of different densities, the wave which produces
the sound will be partly reflected in passing from
one medium to the other, and the direction of the
transmitted wave changed; and hence in passing
through such media different portions of the wave
will reach the ear at different times, and thus destroy the sharpness and distinctness of the sound.
This may be proved by many striking facts. If
we put a bell in a receiver containing a mixture of
hydrogen gas and atmospheric air, the sound of the
bell can scarcely be heard. During a shower of
rain or of snow, noises are greatly deadened, and
when sound is transmitted along an iron wire or
an iron pipe of sufficient length, we actually hear
two sounds, one transmitted more rapidly through
the solid, and the other more slowly through the air.
The same property is well illustrated by an elegant and
easily repeated experiment of Chladni's. When spark-
ling champagne is poured into a tall glass till it is

half full, the glass loses its power of ringing by a
stroke upon its edge, and emits only a disagreeable
and puffy sound.   This effect will continue while
the wine is filled with bubbles of air, or as long as
the effervescence lasts ; but when the effervescence
begins to subside, the sound becomes clearer and
clearer, and the glass rings as usual when the air-
bubbles have vanished.   If we reproduce the effervescence by stirring the champagne with a piece
of bread the glass will again cease to ring.   The
same experiment will succeed with other efferves-
cing fluids.

The difference in the audibility of sounds that
pass over homogeneous and over mixed media is
sometimes so remarkable as to astonish those who
witness it.   The following fact is given on the evidence of an officer who observed it : When the
British and the American forces were encamped on
each side of a river, the outposts were so near that
the form of individuals could be easily distinguished.   An American drummer made his appearance
and began to beat his drum, but though the motion of his arms were distinctly seen, not a single
sound reached the ear of the observer.   A coating
of snow that had newly fallen upon the ground, and
the thickness of the atmosphere, had conspired to
obstruct the sound.   An effect the very reverse of
this is produced by a coating of glazed or hardened
snow, or by an extended surface of ice or water.
Lieutenant Foster was able to carry on a conversation with a sailor across Port Bowen harbour, a
distance of no less than a mile and a quarter, and
the sound of great guns has been heard at distances varying from 120 to 200 miles.   Over hard

and dry ground of an uniform character, or where a thin soil rests upon a continuous stratum of rock, the sound is heard at a great distance, and hence it is the practice among many eastern tribes to ascertain the approach of an enemy by applying the ear to the ground.

Many remarkable phenomena in the natural world are produced by the reflexion and concentration of sound. Every person is familiar with the ordinary *Echo* which arises from the reflexion of sound from an even surface, such as the face of a wall, of a house, of a rock, of a hill, or of a cloud. As sound moves at the rate of 1090 feet in a second, and as the sound which returns to the person who emits it has travelled over a space equal to twice his distance from the reflecting surface, the distance in feet of the body which occasions the echo may be readily found by multiplying 545 by the number of seconds which elapse between the emission of the sound and its return in the form of an echo. This kind of echo, where the same person is the speaker and the hearer, never takes place unless when the observer is immediately in front of the reflecting surface, or when a line drawn from his mouth to the flat surface is nearly perpendicular to it, because in this case alone the wave of sound is reflected in the very same direction from the wall in which it reaches it. If the speaker places himself on one side of this line, then the echo will be heard most distinctly by another person as far on the other side of it, because the waves of sound are reflected like light, so that the angle of incidence or the inclination at which the sound falls upon the reflected surface is equal

to the angle of reflexion, or the inclination at
which the sound is returned from the wall. If two
persons, therefore, are placed before the reflecting
wall, the one will hear the echo of the sound emitted by the other, and obstacles may intervene between these two persons so that neither of them
hears the direct sound emitted by the other; in the
same manner as the same persons similarly placed
before a looking-glass would see each other distinctly by reflexion, though objects might obstruct
their direct view of each other.

Hitherto we have supposed that there is only
one reflecting surface, in which case there will be
only one echo : but if there are several reflecting
surfaces, as is the case in an amphitheatre of mountains, or during a thunder-storm, where there are
several strata or masses of clouds ; or if there are
two parallel or inclined surfaces between which the
sound can be repeatedly reflected, or if the surface is
curved so that the sound reflected from one part
falls upon another part, like the sides of a polygon
inscribed in a circle,—in all these cases there will
be numerous echoes, which produce a very singular effect.   Nothing can be more grand and sublime than the primary and secondary echoes of
a piece of ordnance discharged in an amphitheatre
of precipitous mountains.   The direct or primary echoes from each reflecting surface reach the
ear in succession, according to their different distances, and these are either blended with or succeeded by the secondary echoes, which terminate in a prolonged growl ending in absolute
silence.   Of the same character are the reverberated claps of a thunder-bolt reflected from the

4

surrounding clouds, and dying away in the distance. The echo which is produced by parallel walls is finely illustrated at the Marquis of Simonetta's villa near Milan, which has been described by Addison and Keysler, and which we believe is that described by Mr Southwell in the Philosophical Transactions for 1746. Perpendicular to the main body of this villa there extends two parallel wings about fifty-eight paces distant from each other, and the surfaces of which are unbroken either with doors or windows. The sound of the human voice, or rather a word quickly pronounced, is repeated above forty times, and the report of a pistol from fifty-six to sixty times. The repetitions, however, follow in such rapid succession that it is difficult to reckon them, unless early in the morning before the equal temperature of the atmosphere is disturbed, or in a calm still evening. The echoes appear to be best heard from a window in the main building between the two projecting walls, from which the pistol also is fired. Dr Plot mentions an echo in Woodstock Park which repeats seventeen syllables by day and twenty by night. An echo on the north side of Shipley church in Sussex repeats twenty-one syllables. Sir John Herschel mentions an echo in the Manfroni palace at Venice, where a person standing in the centre of a square room about twenty-five feet high with a concave roof, hears the stamp of his foot repeated a great many times, but as his position deviates from the centre, the echoes become feebler, and at a short distance entirely cease. The same phenomenon, he remarks, occurs in the large room of the library of the museum at Naples. M. Genefay has described as existing near Rouen a curious

oblique echo which is not heard by the person who
emits the sound.   A person who sings hears only
his own voice, while those who listen hear only
the echo, which sometimes seems to approach, and
at other times to recede from the ear; one person
hears a single sound, another several sounds, and
one hears it on the right and another on the left,
the effect always changing, as the hearer changes
his position.   Dr Birch has described an extraordinary echo at Roseneath in Argyleshire, which
certainly does not now exist.   When eight or ten
notes were played upon a trumpet they were correctly repeated, but on a key a third lower.   After
a short pause another repetition of the notes was
heard in a still lower tone, and after another short
interval they were repeated in a still lower tone.

In the same manner as light is always lost by
reflexion, so the waves of sound are enfeebled by
reflexion from ordinary surfaces, and the echo is
in such cases fainter than the original sound.   If
the reflecting surface, however, is circular, sound
may be condensed and rendered stronger in the
same manner as light.   I have seen a fine example of this, in the circular turn of a garden wall
nearly a mile distant from a weir across a river.
When the air is pure and homogeneous, the
rushing sound of the water is reflected from the
hollow surface of the wall, and concentrated in a
focus, the place of which the ear can easily discover
from the intensity of the sound being there a
maximum.   A person not acquainted with the locality conceives that the rushing noise is on the
other side of the wall.

In Whispering Galleries, or places where the lowest whispers are carried to distances at which the direct sound is inaudible, the sound may be conveyed in two ways, either by repeated reflexions from a curved surface in the direction of the sides of a polygon inscribed in a circle, or where the whisperer is in the focus of one reflecting surface, and the hearer in the focus of another reflecting surface, which is placed so as to receive the reflected sounds. The first of these ways is exemplified in the whispering gallery of St Paul's, and in the octagonal gallery of Gloucester Cathedral, which conveys a whisper 75 feet across the nave, and the second in the baptistery of a church in Pisa, where the architect Giovanni Pisano is said to have constructed the cupola on purpose. The cupola has an elliptical form, and when one person whispers in one focus, it is distinctly heard by the person placed in the other focus, but not by those who are placed between them. The sound first reflected passes across the cupola, and enters the ears of the intermediate persons, but it is too feeble to be heard, till it has been condensed by a second reflexion to the other focus of ellipse. A naval officer, who travelled through Sicily in the year 1824, gives an account of a powerful whispering place in the cathedral of Girgenti, where the slightest whisper is carried with perfect distinctness through a distance of 250 feet, from the great western door to the cornice behind the high altar. By an unfortunate coincidence the focus of one of the reflecting surfaces was chosen for the place of the confessional, and when this was accidentally discovered, the lovers of secrets resorted to the other focus, and thus became acquainted with confessions of the

P

gravest import. This divulgence of scandal continued for a considerable time, till the eager curiosity of one of the dilettanti was punished, by hearing his wife's avowal of her own infidelity. This circumstance gave publicity to the whispering peculiarity of the cathedral, and the confessional was removed to a place of greater secrecy.

An echo of a very peculiar character has been described by Sir John Herschel in his Treatise on Sound, as produced by the suspension bridge across the Menai strait in Wales. " The sound of a blow with a hammer," says he, " on one of the main piers, is returned in succession from each of the cross beams which support the road-way, and from the opposite pier at a distance of 576 feet ; and in addition to this, the sound is many times repeated between the water and the road-way. The effect is a series of sounds which may be thus written : the first re-

Fig. 50.

turn is sharp and strong from the road-way overhead ; the rattling which succeeds dies away rapidly, but the single repercussion from the opposite pier is very strong, and is succeeded by a faint palpitation repeating the sound at the rate of twenty-eight times in five seconds, and which, therefore, corresponds to a distance of 184 feet, or very nearly the double interval from the road-way to the water. Thus it appears that in the repercussion

between the water and road-way that from the latter only affects the ear, the line drawn from the
auditor to the water being too oblique for the
sound to diverge sufficiently in that direction. Another peculiarity deserves especial notice, namely,
that the echo from the opposite pier is best heard
when the auditor stands precisely opposite to the
middle of the breadth of the pier, and strikes just
on that point. As it deviates to one or the other
side, the return is proportionally fainter, and is
scarcely heard by him when his station is a little
beyond the extreme edge of the pier, though another person, stationed (on the same side of the
water) at an equal distance from the central point,
so as to have the pier between them, hears it well."

A remarkable subterranean echo is often heard
when the hoofs of a horse or the wheels of a carriage
pass over particular spots of ground. This sound
is frequently very similar to that which is produced in passing over an arch or vault, and is commonly attributed to the existence of natural or artificial caves beneath. As such caves have often
been constructed in times of war as places of securi-
ty for persons and property, many unavailing attempts have been made to discover hidden treasures where their locality seemed to be indicated
by subterraneous sounds. But though these sounds
are sometimes produced by excavations in the
ground, yet they generally arise from the nature
of the materials of which the ground is composed,
and from their manner of combination. If the hollow of a road has been filled up with broken rock,
or with large water-worn stones, having hollows either left entirely empty, or filled up with materials

of different density, then the sound will be reflected in passing from the loose to the dense materials,
and there will arise a great number of echoes reaching the ear in rapid succession, and forming by
their union a hollow rumbling sound. This principle has been very successfully applied by Sir John
Herschel to explain the subterranean sounds with
which every traveller is familiar who has visited
the Solfaterra near Naples. When the ground at
a particular place is struck violently by throwing a
large stone against it, a peculiar hollow sound is
distinctly heard. This sound has been ascribed by
some geologists to the existence of a great vault
communicating with the ancient seat of the volcano, by other writers to a reverberation from the
surrounding hills with which it is nearly concentric, and by others to the porosity of the ground.
Dr Daubeny, who says that the hollow sound is
heard when any part of the Solfaterra is struck, accounts for it by supposing that the hill is not
made up of one entire rock, but of a number of detached blocks, which, hanging as it were by each
other, form a sort of vault over the abyss within
which the volcanic operations are going on.* Mr
Forbes, who has given the latest and most interesting description of this singular volcano,† agrees in
opinion with Dr Daubeny, while Mr Scrope‡ and
Sir John Herschel concur in opinion that no such
cavities exist. " It seems most probable," says the
latter," that the hollow reverberation is nothing more

* Description of Volcanoes, p. 170.
† Edinburgh Journal of Science, N. Series, No. i. p. 124.
‡ Considerations on Volcanoes, and *Edinburgh Journal
of Science*, No. xx. p. 261, and No. xiv. p. 265.

than an assemblage of partial echoes arising from
the reflexion of successive portions of the original
sound in its progress through the soil at the innumerable half-coherent surfaces composing it :
were the whole soil a mass of sand, these reflexions would be so strong and frequent as to destroy
the whole impulse, in too short an interval to allow of a distinguishable after sound. It is a case
analogous to that of a strong light, thrown into a
milky medium or smoky atmosphere ; the whole
medium appears to shine with a nebulous undefin-
ed light. This is to the eye what such a hollow
sound is to the ear." *

It has been recently shown by M. Savart, that
the human ear is so extremely sensible as to be capable of appreciating sounds which arise from about
twenty-four thousand vibrations in a second, and
consequently, that it can hear a sound which lasts
only the twenty-four thousandth part of a second.
Vibrations of such frequency afford only a shrill
squeak or chirp ; and Dr Wollaston has shown that
there are many individuals with their sense of hearing entire, who are altogether insensible to such
acute sounds, though others are painfully affected by
them. Nothing, as Sir John Herschel remarks, can
be more surprising than to see two persons, neither
of them deaf, the one complaining of the penetrat-
ing shrillness of a sound, while the other maintains
there is no sound at all. Dr Wollaston has also
shown that this is true also of very grave sounds,
so that the hearing or not hearing of musical notes
at both extremities of the scale seems to depend
wholly on the pitch or frequency of vibration con-

* Art. *Sound*, Encycl. Metrop. § 110.

stituting the note, and not upon the intensity or loudness of the noise. This affection of the ear sometimes appears in cases of common deafness, where a shrill tone of voice, such as that of women and children, is often better heard than the loud and deeper tone of men.

Dr Wollaston remarked, that when the mouth and nose are shut, the tympanum or drum of the ear may be so exhausted by a forcible attempt to take breath by the expansion of the chest, the pressure of the external air upon the membrane gives it such a tension that the ear becomes insensible to grave tones, without losing in any degree the perception of sharper sounds. Dr Wollaston found, that after he had got into the habit of making the experiment, so as to be able to produce a great degree of exhaustion, his ears were insensible to all sounds below F, marked by the base cliff. " If I strike the table before me," says he, "with the end of my finger, the whole board sounds with a deep dull note. If I strike it with my nail, there is also at the same time a sharp sound produced by quicker vibrations of parts around the point of contact. When the ear is exhausted it hears only the latter sound, without perceiving in any degree the deeper note of the whole table. In the same manner, in listening to the sound of a carriage, the deeper rumbling noise of the body is no longer heard by an exhausted ear ; but the rattle of a chain or loose screw remains at least as audible as before exhaustion." Dr Wollaston supposes that this excessive tension of the drum of the ear, when produced by the compressed air in the diving-bell, will also produce a corresponding *deafness to low tones.*

This curious experiment has been since made, by Dr Colladon when descending in the diving-bell at Howth in 1820. "We descended," says he, "so slowly that we did not notice the motion of the bell; but as soon as the bell was immersed in water, we felt about the ears and the forehead a sense of pressure, which continued increasing during some minutes. I did not, however, experience any pain in the ears; but my companion suffered so much that we were obliged to stop our descent for a short time. To remedy that inconvenience, the workmen instructed us, after having closed our nostrils and mouth, to endeavour to swallow, and to restrain our respiration for some moments, in order that, by this exertion, the internal air might act on the Eustachian tube. My companion, however, having tried it, found himself very little relieved by this remedy. After some minutes, we resumed our descent. My friend suffered considerably; he was pale; his lips were totally discoloured; his appearance was that of a man on the point of fainting; he was in involuntary low spirits, owing, perhaps, to the violence of the pain, added to that kind of apprehension which our situation unavoidably inspired. This appeared to me the more remarkable, as my case was totally the reverse. I was in a state of excitement resembling the effect of some spirituous liquor. I suffered no pain; I experienced only a strong pressure round my head, as if an iron circle had been bound about it. I spoke with the workmen and had some difficulty in hearing them. This difficulty of hearing rose to such a height, that during three or four minutes I could not hear them speak. I could not, indeed, hear myself speak, though I spoke as loudly as possible; nor did even the great noise caused by the violence of the current against the sides of the bell reach my ears."

The effect thus described by Dr Colladon is different from that anticipated by Dr Wollaston. He was not merely deaf to low tones but to all sounds whatever; and I have found by repeated experiment, that my own ears become perfectly insensible even to the shrill tones of the female voice, and of the voice of a child, when the drum of the ear is thrown into a state of tension by yawning.

With regard to sounds of high pitch at the other extremity of the scale, Dr Wollaston has met with persons, whose hearing was in other respects perfect, who never heard the chirping of the *Gryllus campestris*, which commonly occurs in hedges during a summer's evening, or that of the house-cricket, or the squeak of the bat, or the chirping of the common house-sparrow. The note of the bat is a full octave higher than that of the sparrow; and Dr Wollaston believes that the note of some insects may reach one octave more, as there are sounds decidedly higher than that of a small pipe, one-fourth of an inch in length, which he conceives cannot be far from six octaves above the middle E of the piano-forte. " The suddenness of the transition," says Dr Wollaston, "from perfect hearing to total want of perception, occasions a degree of surprise, which renders an experiment on this subject with a series of small pipes among several persons rather amusing. It is curious to observe the change of feeling manifested by various individuals of the party, in succession, as the sounds approach and

pass the limits of their hearing.   Those who enjoy
a temporary triumph are often compelled in their
turn to acknowledge to how short a distance their
little superiority extends."   In concluding his
interesting paper on this subject, Dr Wollaston
conjectures that animals, like the grylli, (whose
powers of hearing appear to commence nearly
where ours terminate,) may have the power of
hearing still sharper sounds which at present we
do not know to exist, and that there may be other
insects having nothing in common with us, but
who are endowed with a power of exciting, and a
sense of perceiving vibrations which makes no impression upon our organs, while their organs are
equally insensible to the slower vibrations to which
we are accustomed.

With the view of studying the class of sounds
inaudible to certain ears, we would recommend it to
the young naturalist to examine the sounds emitted by the insect tribe, both in relation to their effect upon the human ear, and to the mechanism by
which they are produced.   The Cicadæ or locusts
in North America appear, from the observations of
Dr Hildreth,* to be furnished with a bag-pipe on
which they play a variety of notes.   " When any
one passes," says he, " they make a great noise and
screaming with their air bladder or bagpipes.   These
bags are placed under, and rather behind, the wings
in the axilla, something in the manner of using
the bagpipes with the bags under the arms,—I could
compare them to nothing else ; and, indeed, I sus-
pect the first inventor of the instrument borrowed
his ideas from some insect of this kind.   They play

* *Edinburgh Journal of Science,* No. xvii. p. 158.

a variety of notes and sounds, one of which nearly imitates the scream of the tree toad."

Among the acoustic wonders of the natural world may be ranked the vocal powers of the statue of Memnon, the son of Aurora, which modern discoveries have withdrawn from among the fables of ancient Egypt. The history of this remarkable statue is involved in much obscurity. Although Strabo affirms that it was overturned by an earthquake, yet as Egypt exhibits no traces of such a convulsion, it has been generally believed that the statue was mutilated by Cambyses. Ph. Casselius, in his dissertation on vocal or speaking stones, quotes the remark of the scholiast in Juvenal, "that, when mutilated by Cambyses, the statue which saluted both the sun and the king, afterwards saluted only the sun." Philostratus, in his life of Apollo, informs us, that the statue looked to the east, and that it spoke as soon as the rays of the rising sun fell upon its mouth. Pausanias, who saw the statue in its dismantled state, says that it is a statue of the sun, that the Egyptians call it Phamenophis, and not Memnon, and *that it emits sounds every morning at sunrise, which can be compared only to that of the breaking of the string of a lyre.* Strabo speaks only of a single sound which he heard ; but Juvenal, who had probably heard it often during his stay in Egypt, describes it as if it emitted several sounds.

Dimidio magicæ resonant ubi Memnone chordæ.

Where broken Memnon sounds his magic strings.

The simple sounds which issued from the statue were in the progress of time magnified into intelligible words, and even into an oracle of seven

verses, and this prodigy has been recorded in a
Greek inscription on the left leg of the statue. But
though this new faculty of the colossus was evidently the contrivance of the Egyptian priests, yet
we are not entitled from this to call in question
the simple and perfectly credible fact that it emitted sounds. This property, indeed, it seems to
possess at the present day; for we learn,* that an
English traveller, Sir A. Smith, accompanied with
a numerous escort, examined the statue, and that
at six o'clock in the morning he heard very distinctly the sounds which had been so celebrated in
antiquity. He asserts that this sound does not
proceed from the statue but from the pedestal; and
he expresses his belief that it arises from the impulse of the air upon the stones of the pedestal,
which are arranged so as to produce this surprising
effect.—This singular description is to a certain extent confirmed by the description of Strabo, who
says, that he was quite certain that he heard a sound
which proceeded either *from the base,* or from the
colossus, or from some one of the assistants. As
there were no Egyptian priests in the escort of Sir
A. Smith, we may now safely reject this last, and,
for many centuries, the most probable hypothesis.

The explanation suggested by Sir A. Smith had
been previously given in a more specific form by
M. Dussaulx, the translator of Juvenal. " The
statue," says he, " being hollow, the heat of the sun
heated the air which it contained, and this air, issuing at some crevice, produced the sounds of which
the priests gave their own interpretation."

Rejecting this explanation, M. Langles, in his dis-

* Revue Encyclopedique, 1821, Tom. ix. p. 592.

sertation on the vocal statue of Memnon, and M.
Salverte, in his work on the occult sciences, have ascribed the sounds entirely to Egyptian priestcraft,
and have even gone so far as to describe the mechanism by which the statue not only emitted
sounds, but articulated distinctly the intonations
appropriate to the seven Egyptian vowels, and consecrated to the seven planets. M. Langles conceives
that the sounds may be produced by a series of
hammers, which strike either the granite itself, or
sonorous stones like those which have been long used
in China for musical instruments. M. Salverte im-
proves this imperfect apparatus, by supposing that
there might be adapted to these hammers, a clepsydra
or water-clock, or any other instrument fitted to
measure time, and so constructed as to put the
hammers in motion at sunrise. Not satisfied with
this supposition, he conjectures that the spring of all
this mechanism was to be found in the art of con-
centrating the rays of the sun, which was well known
to the ancients. Between the lips of the statue, or
in some less remarkable part of it concealed from
view by its height, he conceives an aperture to be
perforated, containing a lens or a mirror capable of
condensing the rays of the rising sun upon one or
more metallic levers which by their expansion put
in motion the seven hammers in succession. Hence
he explains why the sounds were emitted only at
sunrise, and when the solar rays fell upon the mouth
of the statue, and why they were never again heard
till the sun returned to the eastern horizon. As
a piece of mechanism, this contrivance is defective
in not providing for the change in the sun's ampli-
tude, which is very considerable even in Egypt, for

as the statue and the lens are both fixed, and as the sounds were heard at all seasons of the year, the same lens which threw the midsummer rays of the sun upon the hammers could not possibly throw upon them his rays in winter. But even if the machinery were perfect, it is obvious that it could not have survived the mutilation of the statue, and could not, short of a miracle, have performed its part in the time of Sir A. Smith.

If we abandon the idea of the whole being a trick of the priesthood, which has been generally done, and which the recent observations of Sir A. Smith authorizes us to do, we must seek some natural cause for the phenomena similar to that suggested by Dussaulx. It is curious to observe how the study of nature gradually dispels the consecrated delusions of ages, and reduces to the level of ordinary facts what time had invested with all the characters of the supernatural: And in the present case it is no less remarkable that the problem of the statue of Memnon should have been first solved by means of an observation made by a solitary traveller wandering on the banks of the Orinoco. " The granitic rock," says Baron Humboldt, " on which we lay, is one of those where travellers on the Orinoco have heard from time to time, towards sunrise, subterraneous sounds resembling those of the organ. The missionaries call these stones *loxas de musica.* ' It is witchcraft,' said our young Indian pilot. We never ourselves heard these mysterious sounds either at Carichana Vieja or in the upper Orinoco; but from information given us by witnesses worthy of belief, the existence of a phenomenon that seems to depend on a certain state of the atmosphere cannot be de-

nied. The shelves of rock are full of very narrow and
deep crevices. They are heated during the day to
about 50°. I often found their temperature at the
surface during the night at 39°, the surrounding
atmosphere being at 28°. It may easily be conceived that the difference of temperature between
the subterraneous and the external air attains its
maximum about sunrise, or at that moment which
is at the same time farther from the period of the
maximum of the heat of the preceding day. May
not these sounds of an organ, then, which are heard
when a person sleeps upon the rock, his ear in contact with the stone, be the effect of a current of air
that issues out through the crevices? Does not the
impulse of the air against the elastic spangles of
mica that intercept the crevices contribute to modify the sounds? May we not admit that the ancient inhabitants of Egypt, in passing incessantly
up and down the Nile, had made the same observation on some rock of the Thebaid, and that the
music of the rocks there led to the jugglery of the
priests in the statue of Memnon?"

This curious case of the production of sounds
in granite rocks at sunrise might have been regarded as a transatlantic wonder which was not applicable to Egypt; but by a singular coincidence of
observation, MM. Jomard, Jollois, and Devilliers,
who were travelling in Egypt nearly about the
same time that M. Humboldt was traversing the
wilds of South America, heard *at sunrise, in a
monument of granite*, situated near the centre of
the spot on which the palace of Carnac stands, *a
noise resembling that of a breaking string*, the
very expression by which Pausanias characterizes

the sound in the Memnonian granite. The travellers regarded these sounds as arising from the transmission of rarefied air through the crevices of a sonorous stone, and they were of the same opinion with Humboldt, that these sounds might have *suggested* to the Egyptian priests *the juggleries of the Memnonium.* Is it not strange that the Prussian and the French travellers should not have gone a step farther, and solved the problem of two thousand years, by maintaining that the sound of the statue of Memnon was itself a natural phenomenon, or a granite sound elicited at sunrise by the very same causes which operated on the Orinoco and in the Temple of Carnac, in place of regarding it as a trick in imitation of natural sounds? If, as Humboldt supposes, the ancient inhabitants of Egypt had, in passing incessantly up and down the Nile, become familiar with the music of the granite rocks of the Thebaid, how could the imitation of such natural and familiar sounds be regarded by the priests as a means of deceiving the people? There could be nothing marvellous in a colossal statue of granite giving out the very same sounds that were given out at the same time of the day by a granite rock; and in place of reckoning it a supernatural fact, they could regard it in no other light than as the duplicate of a well known natural phenomenon. It is a mere conjecture, however, that such sounds were common in the Thebaid, and it is therefore probable that a granite rock, possessing the property of emitting sounds at sunrise, had been discovered by the priests, who were at the same time the philosophers of Egypt, and that the block had been employed in the formation of the Memnonian statue for the purpose of impressing upon it a supernatural character, and enabling them to maintain their influence over a credulous people.

The inquiries of recent travellers have enabled us to corroborate these views, and to add another remarkable example of the influence of subterraneous sounds over superstitious minds. About three leagues to the north of Tor in Arabia Petræa, is a mountain, within the bosom of which the most singular sounds have been heard. The Arabs of the Desert ascribe these sounds to a convent of monks preserved miraculously under ground and the sound is supposed to be that of the *Nakous*, a long narrow metallic ruler suspended horizontally, which the priest strikes with a hammer for the purpose of assembling the monks to prayer. A Greek was said to have seen the mountain open, and to have descended into the subterranean convent, where he found fine gardens and delicious water ; and, in order to give proof of his descent, he produced some fragments of consecrated bread, which he pretended to have brought from the subterranean convent. The inhabitants of Tor likewise declare that the camels are not only frightened but rendered furious when they hear these subterraneous sounds.

M. Seetzen, the first European traveller who visited this extraordinary mountain, set out from Wodyel Nackel on the 17th of June at five o'clock in the morning. He was accompanied by a Greek Christian and some Bedouin Arabs, and after a quarter of an hour's walk they reached the foot of a majestic rock of hard sandstone. The mountain itself was quite bare and entirely composed of it. He found inscribed upon the rock several

Greek and Arab names, and also some Koptic characters, which proved that it had been resorted to
for centuries. About noon the party reached the
foot of the mountains called *Nakous,* where at the
foot of a ridge they beheld an insulated peaked
rock. This mountain presented upon two of its
sides two sandy declivities about 150 feet high, and
so inclined that the white and slightly adhering
sand which rests upon its surface is scarcely able
to support itself; and when the scorching heat of
the sun destroys its feeble cohesion, or when it is
agitated by the smallest motions, it slides down the
two acclivities. These declivities unite behind the
insulated rock, forming an acute angle, and, like
the adjacent surfaces, they are covered with steep
rocks which consist chiefly of a white and friable
freestone.

The first sound which greeted the ears of the travellers took place at an hour and a quarter after
noon. They had climbed with great difficulty as
far as the sandy declivity, a height of seventy or
eighty feet, and had rested beneath the rocks
where the pilgrims are accustomed to listen to the
sounds.

While in the act of climbing, M. Seetzen heard
the sound from beneath his knees, and hence he
was led to think, that the sliding of the sand was
the cause of the sound, and not the effect of the
vibration which it occasioned. At three o'clock
the sound became louder, and continued six minutes,
and after having ceased for ten minutes, it was
again heard. The sound appeared to have the greatest resemblance to that of the humming top, rising
and falling like that of an Eolian harp. Believing

Q

that he had discovered the true origin of the sound, M. Seetzen was anxious to repeat the experiment, and with this view he climbed with the utmost difficulty to the highest rocks, and, sliding down as fast as he could, he endeavoured, with the help of his hands and feet, to set the sand in motion. The effect thus produced far exceeded his expectations, and the sand in rolling beneath him made so loud a noise that the earth seemed to tremble to such a degree that he states he should certainly have been afraid if he had been ignorant of the cause.

M. Seetzen throws out some conjectures respecting the cause of these sounds. Does the rolling layer of sand, says he, act like the fiddle-bow, which on being rubbed upon a plate of glass raises and distributes into regular figures the sand with which the plate is covered ? Does the adherent and fixed layer of sand perform here the part of the plate of glass, and the neighbouring rocks that of the sounding body ? We cannot pretend to answer these questions, but we trust that some philosopher competent to the task will have an opportunity of examining these interesting phenomena with more attention, and describing them with greater accuracy.

The only person, so far as I can learn, who has visited El Nakous since the time of Seetzen, is Mr Gray of University College Oxford; but he has not added much to the information acquired by his predecessor. During the first visit which he made to the place, he heard at the end of a quarter of an hour a low continuous murmuring sound beneath his feet, which gradually changed into pulsations as it became louder, so as to resemble the striking of a clock, and at the end of five minutes it became so strong as to detach the sand.   Returning to the spot next day, he heard the sound still louder than before.   He could not observe any crevices by which the external air could penetrate, and as the sky was serene and the air calm, he was satisfied that the sounds could not arise from this cause. *

* See *Edinburgh Journal of Science*, No xi. p. 153, and No. xiii. page 51.

LETTER X.

THE mechanical knowledge of the ancients was
principally theoretical, and though they seem to
have executed some minor pieces of mechanism
which were sufficient to delude the ignorant, yet
there is no reason for believing, that they have executed any machinery that was capable of exciting
much surprise, either by its ingenuity or its magnitude. The properties of the mechanical powers,
however, seem to have been successfully employed
in performing feats of strength which were beyond
the reach even of strong men, and which could not
fail to excite the greatest wonder when exhibited
by persons of ordinary size.

Firmus, a native of Seleucia, who was executed
by the Emperor Aurelian for espousing the cause

of Zenobia, was celebrated for his feats of strength. In his account of the life of Firmus, who lived in the third century, Vopiscus informs us, that he could suffer iron to be forged upon an anvil placed upon his breast. In doing this he lay upon his back, and resting his feet and shoulders against some support, his whole body formed an arch as we shall afterwards more particularly explain. Until the end of the sixteenth century the exhibition of such feats does not seem to have been common. About the year 1703, a native of Kent of the name of Joyce exhibited such feats of strength in London and other parts of England, that he received the name of the second Sampson. His own personal strength was very great ; but he had also discovered without the aid of theory, various positions of his body in which men even of common strength could perform very surprising feats. He drew against horses, and raised enormous weights ; but as he actually exhibited his power in ways which evinced the enormous strength of his own muscles, all his feats were ascribed to the same cause. In the course of eight or ten years, however, his methods were discovered, and many individuals of ordinary strength exhibited a number of his principal performances, though in a manner greatly inferior to Joyce.

Sometime afterwards, John Charles Van Eckeberg, a native of Harzgerode in Anhalt, travelled through Europe under the appellation of Sampson, exhibiting very remarkable examples of his strength. This we believe is the same person whose feats are particularly described by Dr Desaguliers. He was a man of the middle size, and of ordinary strength :

and as Dr Desaguliers was convinced that his feats were exhibitions of skill and not of strength, he was desirous of discovering his methods, and with this view he went to see him accompanied with the Marquis of Tullibardine, Dr Alexander Stuart, and Dr Pringle, and his own mechanical operator. They placed themselves round the German so as to be able to observe accurately all that he did, and their success was so great, that they were able to perform most of the feats the same evening by themselves, and almost all the rest when they had provided the proper apparatus. Dr Desaguliers exhibited some of the experiments before the Royal Society, and has given such a distinct explanation of the principles on which they depend, that we shall endeavour to give a popular account of them.

1. The performer sat upon an inclined board A B placed upon a frame C D E, with his feet abutting

Fig. 51.

against the upright board C. Round his loins was placed a strong girdle F G, to the iron ring of which at G was fastened a rope by means of a hook. The rope passed between his legs through a hole in the board C, and several men, or two horses pulling at the other end of the rope, were unable to draw the performer out of his place. His hands at G, seemed to pull against the men, but they were of no advantage to him whatever.

2. Another of the German's feats is shown in Fig. 52. Having fixed the rope above-mentioned

Fig. 52.

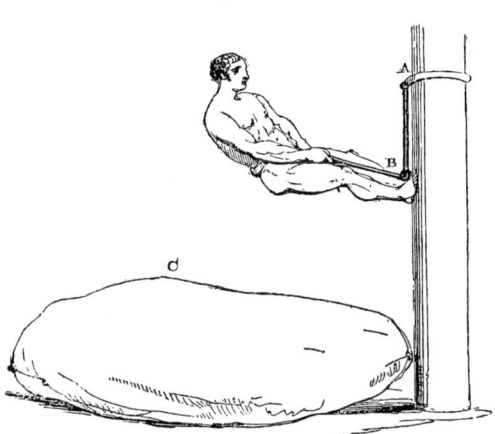

to a strong post at A, and made it pass through a fixed iron eye at B, to the ring in his girdle, he planted his feet against the post at B, and raised himself from the ground by the rope, as shown in the figure. He then suddenly stretched out his legs

and broke the rope, falling back on a feather bed
at C, spread out to receive him.

3. In imitation of Firmus, he laid himself down
on the ground, as shown in Fig. 53, and when an

Fig. 53.

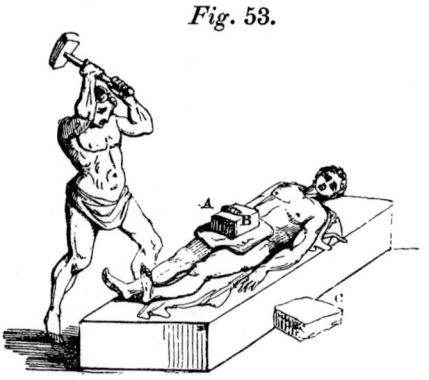

anvil A was placed upon his breast, a man hammered with all his force the piece of iron B, with a
sledge hammer, and sometimes two smiths cut in two
with chisels a great cold bar of iron laid upon the
anvil. At other times a stone of huge dimensions,
half of which is shown at C, was laid upon his
belly, and broken with a blow of the great hammer.

4. The performer then placed his shoulders upon one chair, and his heels upon another as in
Fig. 54, forming with his back-bone, thighs, and
legs, an arch springing from its abutments at A
and B. One or two men then stood upon his belly,
rising up and down while the performer breathed.
A stone one and a half feet long, one foot broad,

and half a foot thick, was then laid upon his belly
and broken by a sledge hammer, an operation which

Fig. 54.

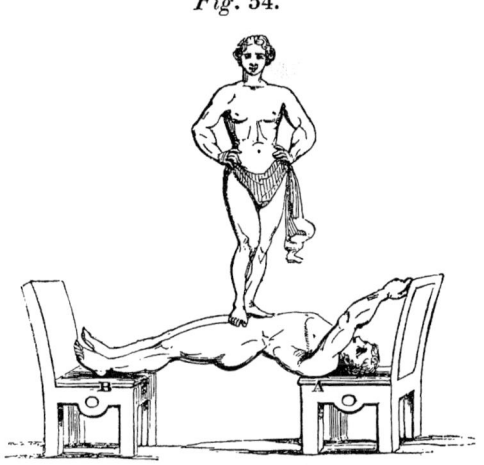

may be performed with much less danger than
when his back touched the ground as in Fig. 53.

5. His next feat was to lie down on the ground
as in Fig. 55 : A man being then placed on his
knees, he draws his heels towards his body, and
raising his knees he lifts up the man gradually, till
having brought his knees perpendicularly under
him as in Fig. 56, he raises his own body up, and
placing his arms around the man's legs, he rises with
him, and sets him down on some low table or eminence of the same height as his knees.   This feat
he sometimes performed with two men in place of
one.

Fig. 55.

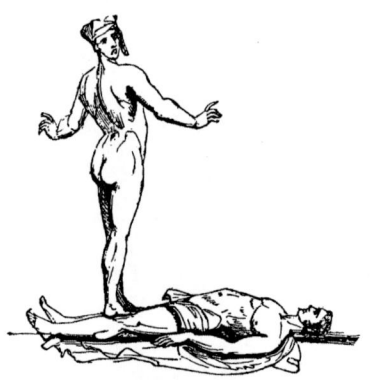

Fig. 56.

6. The last, and apparently the most wonderful, performance of the German is shown in Fig. 57,

Fig. 57.

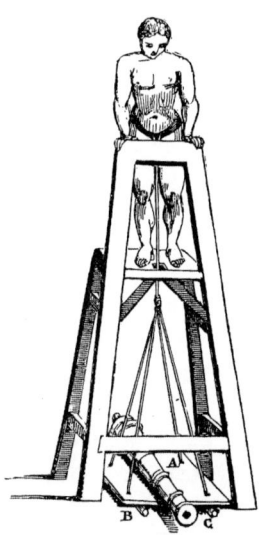

where he appears to raise a cannon A, placed upon a scale, the four ropes of the scale being fixed to a rope or chain attached to his girdle in the manner already described.   Previous to the fixing of the ropes the cannon and scale rest upon two rollers B, C, but when all is ready, the two rollers are knocked from beneath the scale, and the cannon is sustained by the strength of his loins.

The German also exhibited his strength in twist-
ing into a screw a flat piece of iron like A, Fig. 58.

Fig. 58.

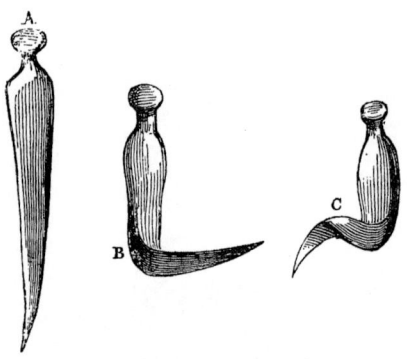

He first bent the iron into a right angle as at B,
and then wrapping his handkerchief about its broad
upper end, he held that end in his left hand, and
with his right applied to the other end, twisted
about the angular point, as shown at C. Lord Tullibardine succeeded in doing the same thing, and even
untwisted one of the irons which the German had
twisted.

It would lead into details by no means popular,
were I to give a minute explanation of the mechanical principles upon which these feats depend. A
few general observations will perhaps be sufficient
for ordinary readers.   The feats No. 1, 2, and 6,
depend entirely on the natural strength of the bones

of the pelvis, which form a double arch, which it
would require an immense force to break,' by any
external pressure directed to the centre of the arch;
and as the legs and thighs are capable of sustaining
four or five thousand pounds when they stand quite
upright, the performer has no difficulty in resisting the force of two horses, or of sustaining the
weight of a cannon weighing two or three thousand pounds.

The feat of the anvil is certainly a very surprising one. The difficulty, however, really consists
in sustaining the anvil, for when this is done the
effect of the hammering is nothing. If the anvil
were a thin piece of iron, or even two or three
times heavier than the hammer, the performer
would be killed by a few blows ; but the blows are
scarcely felt when the anvil is very heavy, for the
more matter the anvil has, the greater is its inertia,
and it is the less liable to be struck out of its place;
for when it has received by the blow the whole mo-
mentum of the hammer, its velocity will be so much
less than that of the hammer, as its quantity of
matter is greater. When the blow, indeed, is struck,
the man feels less of the weight of the anvil than
he did before, because in the reaction of the stone
all the parts of it round about the hammer rise
towards the blow. This property is illustrated by
the well known experiment of laying a stick with
its ends upon two drinking glasses full of water,
and striking the stick downwards in the middle
with an iron bar. The stick will in this case be
broken without breaking the glasses, or spilling
the water. But if the stick is struck upwards as
if to throw it up in the air, the glasses will break

if the blow be strong, and if the blow is not very quick, the water will be spilt without breaking the glasses.

When the performer supports a man upon his belly as in Fig. 54, he does it by means of the strong arch formed by his backbone, and the bones of his legs and thighs. If there were room for them he could bear three or four, or, in their stead, a great stone, to be broken with one blow.

A number of feats of real and extraordinary strength were exhibited about a century ago, in London, by Thomas Topham, who was five feet ten inches high, and about 31 years of age. He was entirely ignorant of any of the methods for making his strength appear more surprising, and he often performed by his own natural powers what he learned had been done by others by artificial means. A distressing example of this occurred in his attempt to imitate the feat of the German Sampson by pulling against horses. Ignorant of the method which we have already described, he seated himself on the ground with his feet against two stirrups, and by the weight of his body he succeeded in pulling against a single horse ; but in attempting to pull against two horses, he was lifted out of his place and one of his knees was shattered against the stirrups, so as to deprive him of most of the strength of one of his legs. The following are the feats of real strength which Dr Desaguliers saw him perform.

1. Having rubbed his fingers with coal ashes to keep them from slipping, he rolled up a very strong and large pewter plate.

2. Having laid seven or eight short and strong

3

pieces of tobacco-pipe on the first and third finger, he broke them by the force of his middle finger.

3. He broke the bowl of a strong tobacco-pipe placed between his first and third finger, by pressing his fingers together sideways.

4. Having thrust such another bowl under his garter, his legs being bent, he broke it to pieces by the tendons of his hams, without altering the bending of his leg.

5. He lifted with his teeth, and held in a horizontal position for a considerable time, a table six feet long, with half a hundred weight hanging at the end of it. The feet of the table rested against his knees.

6. Holding in his right hand an iron kitchen poker three feet long and three inches round, he struck upon his bare left arm, between the elbow and the wrist till he bent the poker nearly to a right angle.

7. Taking a similar poker, and holding the ends of it in his hands, and the middle against the back of his neck, he brought both ends of it together before him, and he then pulled it almost straight again. This last feat was the most difficult, because the muscles which separate the arms horizontally from each other are not so strong as those which bring them together.

8. He broke a rope about two inches in circumference, which was partly wound about a cylinder four inches in diameter, having fastened the other end of it to straps that went over his shoulder.

9. Dr Desaguliers saw him lift a rolling stone of about 800 ℔s. weight with his hands only, standing in a frame above it, and taking hold of a frame fast-

ened to it. Hence Dr Desaguliers gives the following relative view of the strengths of individuals.

    Strength of the weakest men -   125 lbs.
    Strength of very strong men -   400
    Strength of Topham -   -   800

The weight of Topham was about 200.

One of the most remarkable and inexplicable experiments relative to the strength of the human
frame, which you have yourself seen and admired,
is that in which a heavy man is raised with the
greatest facility, when he is lifted up the instant
that his own lungs and those of the persons who
raise him are inflated with air. This experiment
was, I believe, first shown in England a few years
ago by Major H., who saw it performed in a large
party at Venice under the direction of an officer of
the American Navy. As Major H. performed it
more than once in my presence, I shall describe as
nearly as possible the method which he prescribed.
The heaviest person in the party lies down upon
two chairs, his legs being supported by the one
and his back by the other. Four persons, one at
each leg, and one at each shoulder, then try to raise
him, and they find his dead weight to be very great,
from the difficulty they experience in supporting
him. When he is replaced in the chair, each of the
four persons takes hold of the body as before, and
the person to be lifted gives two signals by clap-
ping his hands. At the first signal he himself and
the four lifters begin to draw a long and full breath,
and when the inhalation is completed, or the lungs
filled, the second signal is given, for raising the
person from the chair. To his own surprise and
that of his bearers, he rises with the greatest facili-

ty, as if he were no heavier than a feather. On several occasions I have observed that when one of the bearers performs his part ill, by making the inhalation out of time, the part of the body which he tries to raise is left as it were behind. As you have repeatedly seen this experiment, and have performed the part both of the load and of the bearer, you can testify how remarkable the effects appear to all parties, and how complete is the conviction, either that the load has been lightened, or the bearer strengthened by the prescribed process.

At Venice the experiment was performed in a much more imposing manner. The heaviest man in the party was raised and sustained upon the points of the fore-fingers of six persons. Major H. declared that the experiment would not succeed if the person lifted were placed upon a board, and the strength of the individuals applied to the board. He conceived it necessary that the bearers should communicate directly with the body to be raised. I have not had an opportunity of making any experiments relative to these curious facts; but whether the general effect is an illusion, or the result of known or of new principles, the subject merits a careful investigation.

Among the remarkable exhibitions of mechanical strength and dexterity, we may enumerate that of supporting pyramids of men. This exhibition is a very ancient one. It is described, though not very clearly, by the Roman poet Claudian, and it has derived some importance in modern times, in consequence of its having been performed in vari-

R

ous parts of Great Britain by the celebrated traveller Belzoni, before he entered upon the more esti-
mable career of an explorer of Egyptian antiquities.
The simplest form of this feat consists in placing
a number of men upon each other's shoulders, so
that each row consists of a man fewer till they form
a pyramid terminating in a single person, upon
whose head a boy is sometimes placed with his
feet upwards.

Among the displays of mechanical dexterity,
though not grounded on any scientific principle,
may be mentioned the art of walking along the
ceiling of an apartment with the head downwards.
This exhibition, which we have witnessed in one of
the London Theatres, never failed to excite the
wonder of the audience, although the movements
of the inverted performer were not such as to inspire us with any high ideas of the mechanism by
which they were effected. The following was probably the method by which the performer was carried along the ceiling. Two parallel grooves or
openings were made in the ceiling at the same distance as the foot tracks of a person walking on
sand. These grooves were narrower than the human
foot, so as to permit a rope or chain or strong wire,
attached to the feet of the performer, to pass
through the ceiling, where they were held by two
or more persons above it. In this way the inverted performer might be carried along by a sliding or
shuffling motion, similar to that which is adopted
in walking in the dark, and in which the feet are
not lifted from the ground. A more regular motion, however, might be produced by a contrivance

for attaching the rope or chain to the sole of the
foot, at each step, and subsequently detaching it.
In this way, when the performer is pulled against
the ceiling by his left foot, he would lift his right
foot, and having made a step with it, and planted it
against the grooves, the rope would be attached to
it, and when the rope was detached from the left
foot, it would make a similar step, while the right
foot was pulled against the ceiling. These effects
might be facilitated and rendered more natural, by
attaching to the body or to the feet of the performer strong wires invisible to the audience, and
by using friction wheels, if a sliding motion only
is required.

A more scientific method of walking upon the
ceiling is suggested by those beautiful pneumatic
contrivances by which insects, fishes, and even some
lizards are enabled to support the weight of their
bodies against the force of gravity. The house-fly
is well known to have the power of walking in an
inverted position upon the ceilings of rooms, as
well as upon the smoothest surfaces. In this case
the fly does not rest upon its legs, and must therefore adhere to the ceiling, either by some glutinous
matter upon its feet, or by the aid of some apparatus given it for that purpose. In examining the
foot of the fly with a powerful microscope, it is
found to consist of two concavities, as shewn in Fig.
59 and 60, the first of which is copied from a drawing by G. Adams, published in 1746, and the second
by J. C. Keller, a painter at Nuremberg, who drew
it for a work published in 1766. The author of
this work maintains that these concavities are only

used when the fly moves horizontally, and that,
when it moves perpendicularly or on the ceiling,

Fig. 59.

they are turned up out of the way, and the progres-
sive motion is effected by fixing the claws shewn
in the figure into the irregularities of the surface
upon which the fly moves, whether it is glass, por-
celain, or any other substance.   Sir Everard Home,
however, supposes with great reason, that these
concave surfaces are (like the leathern suckers used
by children for lifting stones) employed to form a
vacuum, so that the foot adheres as it were by suc-

tion to the ceiling, and enables the insect to support itself in an inverted position.

Fig. 60.

This conclusion Sir Everard has been led to draw from an examination of the foot of the Lacerta Gecko. Sir Joseph Banks had mentioned to him in the year 1815, that this lizard, which is a native of the island of Java, comes out in the evening from the roofs of the houses, and walks down the smooth hard polished chunam walls in search of the flies which settle upon them, and which are its natural food. When Sir Joseph was at Batavia, he amused himself in catching this lizard. He stood close to the wall at some distance from the animal, and, by suddenly scraping the wall with a long flattened pole, he was able to bring the animal to the ground.

Having procured from Sir Joseph a very large specimen of the Gecko, which weighed $5\frac{3}{4}$ ounces avoirdupois, Sir Everard Home was enabled to ascertain the peculiar mechanism by which the feet of this animal have the power of keeping hold of a smooth hard perpendicular wall, and carry up so heavy a weight as that of its body.

The foot of the Gecko has five toes, and at the end of each of them, except the thumb, is a very sharp and highly curved claw. On the under surface of each toe, are sixteen transverse slits, leading to as many cavities or pockets, the depth of which is nearly equal to the length of the slit that forms the surface. These cavities all open forwards, and the external edge of each opening is serrated like the teeth of a small toothed comb. The cavities are lined with a cuticle, which also covers the serrated edges.

This structure Sir Everard Home found to bear a considerable resemblance to that portion of the head of the *Echineis remora*, or sucking-fish, by which it attaches itself to the shark, or the bottoms of ships. It is of an oval form, and is surrounded by a broad loose moveable edge, capable of applying itself closely to the surface on which it is set. It consists of two rows of cartilaginous plates, connected by one edge to the surface on which they are placed, the other, or the external edge, being serrated like that in the cavities of the feet of the Gecko. The two rows are separated by a thin ligamentous partition, and the plates being raised or depressed by the voluntary muscles form so many vacua, by means of which the adhesion of the fish is effected.

These beautiful contrivances of Divine Wisdom cannot fail to arrest the attention and excite the admiration of the reader; but though there can be little doubt that they are pneumatic suckers wrought by the voluntary muscles of the animals to which they belong, yet we would recommend the farther examination of them to the attention of those who have good microscopes at their command.

LETTER XI.

Mechanical automata of the ancients—Moving tripods—Auto-
mata of Dædalus—Wooden pigeon of Archytas—Automa-
tic clock of Charlemagne—Automata made by Turrianus
for Charles V.—Camus's automatic carriage made for Louis
XIV.—Degennes's mechanical peacock—Vaucanson's duck
which ate and digested its food—Du Moulin's automata—
Baron Kempelen's automaton chess-player—Drawing and
writing automata—Maillardet's conjuror—Benefits derived
from the passion for automata—Examples of wonderful ma-
chinery for useful purposes—Duncan's tambouring machinery
—Watt's statue turning machinery—Babbage's calculating
machinery.

WE have already seen that the ancients had attained some degree of perfection in the construction of
automata or pieces of mechanism which imitated the
movements of man and the lower animals. The
tripods which Homer * mentions as having been constructed by Vulcan for the banqueting hall of the
gods, advanced of their own accord to the table,
and again returned to their place. Self-moving
tripods are mentioned by Aristotle, and Philostratus
informs us, in his life of Apollonius, that this philosopher saw and admired similar pieces of mechanism among the sages of India.

 Dædalus enjoys also the reputation of having
constructed machines that imitated the motions of

  * *Iliad,* Lib. xviii. 373—378.

the human body. Some of his statues are said to have moved about spontaneously, and Plato, Aristotle, and others have related that it was necessary to tie them, in order to prevent them from running away. Aristotle speaks of a wooden Venus, which moved about in consequence of quicksilver being poured into its interior; but Callistratus, the tutor of Demosthenes, states with some probability, that the statues of Dædalus received their motion from the mechanical powers. Beckmann is of opinion that the statues of Dædalus differed only from those of the early Greeks and Egyptians in having their eyes open and their feet and hands free, and that the reclining posture of some, and the attitude of others, " as if ready to walk," gave rise to the exaggeration that they possessed the power of locomotion. This opinion, however, cannot be maintained with any show of reason; for if we apply such a principle in one case, we must apply it in all, and the mind would be left in a state of utter scepticism respecting the inventions of ancient times.

We are informed by Aulus Gellius, on the authority of Favorinus, that Archytas of Tarentum, who flourished about 400 years before Christ, constructed a wooden pigeon which was capable of flying. Favorinus relates, that when it once alighted, it could not again resume its flight, and Aulus Gellius adds, that it was suspended by balancing, and animated by a concealed aura or spirit.

Among the earliest pieces of modern mechanism was the curious water-clock presented to Charlemagne by the Kaliph Harun al Raschid. In the dial-plate there were twelve small windows corresponding with the divisions of the hours. The hours were indicated by the opening of the windows, which let out little metallic balls, which struck the hour by falling upon a brazen bell. The doors continued open till twelve o'clock, when twelve little knights, mounted on horseback, came out at the same instant, and after parading round the dial, shut all the windows and returned to their apartments.*

The next automata of which any distinct account has been preserved are those of the celebrated John Muller or Regiomontanus, which have been mentioned by Kircher, Baptista Porta, Gassendi, Lana, and Bishop Wilkins. This philosopher is said to have constructed an artificial eagle, which flew to meet the Emperor Maximilian when he arrived at Nuremberg on the 7th June 1470. After soaring aloft in the air, the eagle is stated to have met the Emperor at some distance from the city, and to have returned and perched upon the town gate, where it waited his approach. When the Emperor reached the gate, the eagle stretched out its wings, and saluted him by an inclination of its body. Muller is likewise reported to have constructed an iron fly, which was put in motion by wheel-work, and which flew about and leapt upon the table. At an entertainment given by this philosopher to some of his familiar friends, the fly flew from his hand, and after performing a considerable round, it returned again to the hand of its master.

The Emperor Charles V., after his abdication of the throne, amused himself in his later years with automata of various kinds. The artist whom he employed was Janellus Turrianus of Cremona. It was his custom after dinner to introduce upon the table figures of armed men and horses. Some of these

* Annales Loisiliani. Anno 807.

beat drums, others played upon flutes, while a third
set attacked each other with spears.   Sometimes he
let fly wooden sparrows, which flew back again to
their nest.   He also exhibited corn-mills so extremely small that they could be concealed in a
glove, yet so powerful that they could grind in a
day as much corn as would supply eight men with
food for a day.

The next piece of mechanism of sufficient interest to merit our attention is that which was made
by M. Camus for the amusement of Louis XIV.
when a child.   It consisted of a small coach, which
was drawn by two horses, and which contained the
figure of a lady within, with a footman and page
behind.   When this machine was placed at the extremity of a table of the proper size, the coachman
smacked his whip, and the horses instantly set off,
moving their legs in a natural manner, and drawing
the coach after them : When the coach reached the
opposite edge of the table, it turned sharply at a
right angle, and proceeded along the adjacent edge.
As soon as it arrived opposite the place where the
King sat it stopped ; the page descended and opened the coach door; the lady alighted, and with a curt-
sey presented a petition, which she held in her
hand to the King.   After waiting some time she
again curtsied and re-entered the carriage.   The page
closed the door, and having resumed his place behind, the coachman whipped his horses and drove
on.   The footman who had previously alighted,
ran after the carriage and jumped up behind into
his former place.

Not content with imitating the movements of
animals, the mechanical genius of the 17th and 18th

centuries ventured to perform by wheels and pinions
the functions of vitality.   We are informed by M.
Lobat, that General Degennes, a French officer who
defended the colony of St Christophers against the
English forces, constructed a peacock, which could
walk about as if alive, pick up grains of corn from
the ground, digest them as if they had been submitted to the action of the stomach, and afterwards discharge them in an altered form.   Degennes is
said to have invented various machines of great use
in navigation and gunnery, and to have constructed clocks without weights or springs.

   The automaton of Degennes probably suggested to M. Vaucanson the idea of constructing his
celebrated duck, which excited so much interest
throughout Europe, and which was perhaps the most
wonderful piece of mechanism that was ever made.
Vaucanson's duck exactly resembled the living animal in size and appearance.   It executed accurately
all its movements and gestures, it ate and drank with
avidity, performed all the quick motions of the head
and throat which are peculiar to the living animal,
and like it, it muddled the water which it drank
with its bill.   It produced also the sound of quack-
ing in the most natural manner.   In the anatomi-
cal structure of the duck, the artist exhibited the
highest skill.   Every bone in the real duck had its
representative in the automaton, and its wings were
anatomically exact.   Every cavity, apophysis, and
curvature was imitated, and each bone executed its
proper movements.   When corn was thrown down
before it, the duck stretched out its neck to pick it
up, it swallowed it, digested it, and discharged it, in
a digested condition.   The process of digestion was

effected by chemical solution, and not by trituration, and the food digested in the stomach was conveyed away by tubes to the place of its discharge.

The automata of Vaucanson were imitated by one Du Moulin, a silversmith, who travelled with them through Germany in 1752, and who died at Moscow in 1765. Beckmann informs us that he saw several of them after the machinery had been deranged ; but that the artificial duck, which he regarded as the most ingenious, was still able to eat, drink, and move. Its ribs, which were made of wire, were covered with duck's feathers, and the motion was communicated through the feet of the duck by means of a cylinder and fine chains like that of a watch.

Ingenious as all these machines are, they sink into insignificance when compared with the automaton chess-player, which for a long time astonished and delighted the whole of Europe. In the year 1769, M. Kempelen, a gentleman of Presburg in Hungary, constructed an automaton chess-player, the general appearance of which is shown in the annexed figures. The chess-player is a figure as large as life, clothed in a Turkish dress sitting behind a large square chest or box three feet and a half long, two feet deep, and two and a half high. The machine runs on casters, and is either seen on the floor when the doors of the apartment are thrown open, or is wheeled into the room previous to the commencement of the exhibition. The Turkish chess-player sits on a chair fixed to the square chest : His right arm rests on the table, and in the left he holds a pipe, which is removed during the game, as it is with this hand that he makes the moves. A chess-board, eighteen inches square, and bearing the usual number of pieces is placed before the figure.

The exhibitor then announces to the spectators his intention of shewing them the mechanism of the automaton. For this purpose he unlocks the door A, Fig. 61, and exposes to view a small cupboard lined with black or dark coloured cloth, and containing cylinders, levers, wheels, pinions, and different pieces of machinery, which *have the appearance* of occupying the whole space. He next opens the door B, Fig. 62, at the back of the same cupboard, and holding a lighted candle at the opening, he still farther displays the enclosed machinery to the spectators, placed in front of A, Fig. 61. When the candle is withdrawn, the door B is then locked ; and the exhibitor proceeds to open the drawer G G, Fig. 61, in front of the chest. Out of this drawer he takes a small box of counters, a set of chess-men, and a cushion for the support of the automaton's arm, as if this was the sole object of the drawer. The two front doors C, C, of the large cupboard, Fig.

4

61, are then opened, and at the back-door D of the same cupboard, Fig. 62, the exhibitor applies a lighted candle, as before, for the purpose of shewing its interior, which is lined with dark cloth like the other, and contains only a few pieces of machinery. The chest is now wheeled round, as in Fig. 62: The garments of the figure are lifted up, and the door E in the trunk, and another door F, in the thigh, are opened, the doors B and D having been previously closed. When this exhibition of the interior of the machine is over, the chest is wheeled back into its original position on the floor. The doors A, C, C, in front, and the drawer G, G, are closed and locked, and the exhibitor, after occupying himself for some time at the back of the chest, as if he were adjusting the mechanism, removes the pipe from the hand of the figure, and winds up the machinery.

The automaton is now ready to play, and when an opponent has been found among the company, the figure takes the first move. At every move made by the automaton, the wheels of the machine are heard in action; the figure moves its head, and seems to look over every part of the chess-board. When it gives check to its opponent, it shakes its head *thrice*, and only *twice* when it checks the queen. It likewise shakes its head when a false move is made, replaces the adversary's piece on the square from which it was taken, and takes the next move itself. In general, though not always, the automaton wins the game.

During the progress of the game, the exhibitor often stands near the machine, and winds it up like a clock after it has made ten or twelve moves. At other times he went to a corner of the room, as if it were to consult a small square box, which stood open for this purpose.

The chess-playing machine, as thus described, was exhibited after its completion in Presburg, Vienna, and Paris, to thousands, and in 1783 and 1784 it was exhibited in London and different parts of England, without the secret of its movements having been discovered. Its ingenious inventor, who was a gentleman and a man of education, never pretended that the automaton itself really played the game. On the contrary, he distinctly stated, " that the machine was a *bagatelle*, which was not without merit in point of mechanism, but that the effects of it appeared so marvellous only from the boldness of the conception, and the fortunate choice of the methods adopted for promoting the illusion."

Upon considering the operations of this automaton, it must have been obvious that the game of chess was performed either by a person enclosed in the chest, or by the exhibitor himself. The first of these hypotheses was ingeniously excluded by the display of the interior of the machine, for as every part contained more or less machinery, the spectator invariably concluded that the smallest dwarf could not be accommodated within, and this idea was strengthened by the circumstance, that no person of this description could be discovered in the suite of the exhibitor. Hence the conclusion was drawn, that the exhibitor actuated the machine either by mechanical means conveyed through its feet, or by a magnet concealed in the body of the exhibitor. That mechanical communication was not formed between the exhibitor and the figure, was obvious from the fact, that no such communi-

3

cation was visible, and that it was not necessary to place the machine on any particular part of the floor. Hence the opinion became very prevalent that the agent was a magnet; but even this supposition was excluded, for the exhibitor allowed a strong and well armed loadstone to be placed upon the machine during the progress of the game: Had the moving power been a magnet, the whole action of the machine would have been deranged by the approximation of a loadstone concealed in the pockets of any of the spectators.

As Baron Kempelen himself had admitted that there was an illusion connected with the performance of the automaton, various persons resumed the original conjecture, that it was actuated by a person concealed in its interior, who either played the game of chess himself, or performed the moves which the exhibitor indicated by signals. A Mr J. F. Freyhere of Dresden published a book on the subject in 1789, in which he endeavoured to explain, by coloured plates, how the effect was produced; and he concluded, " that a well-taught boy, very thin and tall of his age, (sufficiently so that he could be concealed in a drawer almost immediately under the chess-board) agitated the whole."

In another pamphlet which had been previously published at Paris in 1785, the author not only supposed that the machine was put in motion by a dwarf, a famous chess-player, but he goes so far as to explain the manner in which he could be accommodated within the machine. The invisibility of the dwarf when the doors were opened was explained by his legs and thighs being concealed in two hollow cylinders, while the rest of his body was out of the box, and hid by the petticoats of the automaton. When the doors were shut the clacks produced by the swivel of a ratchet-wheel permitted the dwarf to change his place and return to the box unheard; and while the machine is wheeled about the room, the dwarf had an opportunity of shutting the trap through which he passed into the machine. The interior of the figure was next shown, and the spectators were satisfied that the box contained no living agent.

Although these views were very plausible, yet they were never generally adopted; and when the automaton was exhibited in Great Britain in 1819 and 1820, by M. Maelzel, it excited as intense an interest as when it was first produced in Germany. There can be little doubt, however, that the secret has been discovered; and an anonymous writer has shown in a pamphlet, entitled " *An attempt to analyse the automaton chess-player* of M. Kempelen," that it is capable of accommodating an ordinarily sized man; and he has explained in the clearest manner how the inclosed player takes all the different positions, and performs all the motions which are necessary to produce the effects actually observed. The following is the substance of his observations :

The drawer G G when closed does not extend to the back of the chest, but leaves a space O, behind it, (See Fig. 69, 70, and 71,) fourteen inches broad, eight inches high, and three feet eleven inches long. This space is never exposed to the view of spectators. The small cupboard seen at A is divided into two parts by a door or screen I, Fig. 68, which is moveable upon a hinge, and is so constructed

that it closes at the same instant that B is closed.
The whole of the front compartment as far as I is
occupied with the machinery H. The other compartment behind I is empty, and communicates
with the space O behind the drawer, the floor of
this division being removed. The back of the great
cupboard C C is double, and the part P Q, to which
the quadrants are attached, moves on a joint Q, at
the upper part, and forms when raised an opening
S, between the two cupboards, by carrying with it
part of the partition R, which consists of cloth tightly stretched. The false back is shown closed in
Fig. 69, while Fig. 70 shows the same back raised,
so as to form the opening S between the chambers.

When the spectator is allowed to look into the
trunk of the figure by lifting up the dress, as in
Fig. 70, it will be observed that a great part of the
space is occupied by an inner trunk N, Fig. 70,
71, which passes off to the back in the form of an
arch, and conceals from the spectators a portion of
the interior. This inner trunk N opens and communicates with the chest by an aperture T, Fig. 72,
about twelve inches broad and fifteen high. When
the false back is raised the two cupboards, the trunk
N, and the space O behind the drawer, are all connected together.

The construction of the interior being thus understood, the chess-player may be introduced into the
chest through the sliding panel U, Fig. 69. He
will then raise the false back of the large cupboard,
and assume the position represented by the shaded
figure in Fig. 63 and 64. Things being in this
state, the exhibitor is ready to begin his process of
deception. He first opens the door A of the small

cupboard, and from the crowded and very ingenious
disposition of the machinery within it, the eye is

Fig. 63. 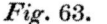   *Fig.* 64.

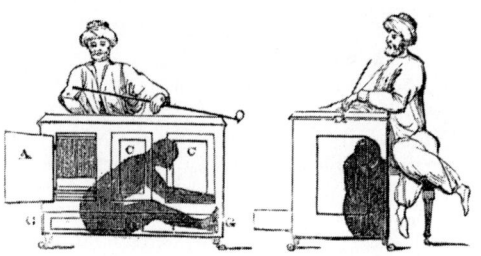

unable to penetrate far beyond the opening, and
the spectator concludes without any hesitation, that
the whole of the cupboard is filled, as it appears to
be, with similar machinery.   This false conclusion
is greatly corroborated by observing the glimmering light which plays among the wheel work when
the door B is opened, and a candle held at the opening.   This mode of exhibiting the interior of the
cupboard satisfies the spectator also that no opaque
body capable of holding or concealing any of the
parts of a hidden agent is interposed between the
light and the observer.   The door B is now locked
and the screen I closed, and as this is done at the
time that the light is withdrawn it will wholly escape observation.

The door B is so constructed as to close by its

own weight, but as the head of the chess-player will soon be placed very near it, the secret would be disclosed if, in turning round, the chest door should by any accident fly open. This accident is prevented by turning the key, and, lest this little circumstance should excite notice, it would probably be regarded as accidental, as the keys were immediately wanted for the other locks.

As soon as the door B is locked, and the screen I closed, the secret is no longer exposed to hazard, and the exhibitor proceeds to lead the minds of the spectators still farther from the real state of things. The door A is left open to confirm the opinion that no person is concealed within, and that nothing can take place in the interior without being observed.

The drawer G G is now opened, apparently for the purpose of looking at the chess-men, cushion and counters which it contains; but the real object of it is to give time to the player to change his

Fig. 65.

position, as shown in the annexed figure, and to replace the false back and partition preparatory to the opening of the great cupboard. The chess-player, as the figure shows, occupies with his body the back compartment of the small cupboard, while his legs and thighs are contained in the space O, behind the drawer G G, his body being concealed by the screen I, and his limbs by the drawer G G.

The great cupboard C C is now opened, and there is so little machinery in it that the eye instantly discovers that no person is concealed in it. To make this more certain, however, a door is opened at the back, and a lighted candle held to it, to allow the spectators to explore every corner and recess.

The front doors of the great and small cupboard being left open, the chest is wheeled round to show the trunk of the figure, and the bunch of keys is allowed to remain in the door D, as the apparent carelessness of such a proceeding will help to remove any suspicion which may have been excited by the locking of the door B.

When the drapery of the figure has been raised, and the doors E and F in the trunk and thigh opened, the chest is wheeled round again into its original position, and the doors E and F closed. In the meantime the player withdraws his legs from behind the drawer, as he cannot so easily do this when the drawer G G is pushed in.

In all these operations, the spectator flatters himself that he has seen in succession every part of the chest, while in reality some parts have been wholly concealed from his view, and others but imperfectly shown, while at the present time nearly half of the chest is excluded from view.

When the drawer G G is pushed in, and the

doors A and C closed, the exhibitor adjusts the
machinery at the back, in order to give time to the
player to take the position shown in a front view
in Fig. 66, and in profile in Fig. 67. In this po-

Fig. 66.   *Fig.* 67.

sition he will experience no difficulty in executing
every movement made by the automaton. As his
head is above the chess-board, he will see through
the waistcoat of the figure, as easily as through a
veil, the whole of the pieces on the board, and he
can easily take up and put down a chess man without any other mechanism than that of a string communicating with the finger of the figure. His
right hand being within the chest may be employed to keep in motion the wheel-work for producing
the noise which is heard during the moves, and to
perform the other movements of the figure, such
as that of moving the head, tapping on the chest,
&c.

A very ingenious contrivance is adopted to facilitate the introduction of the player's left arm into the arm of the figure. To permit this, the arm of the figure requires to be drawn backwards ; and for the purpose of concealing, and at the same time explaining this strained attitude, a pipe is ingeniously placed in the automaton's hand. For this reason the pipe is not removed till all the other arrangements are completed. When every thing has been thus prepared, the pipe is taken from the figure, and the exhibitor winds up as it were the

Fig. 68.   *Fig.* 69.

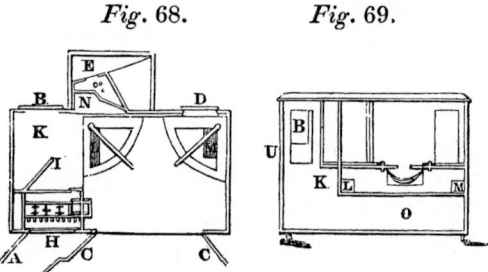

inclosed machinery, for the double purpose of impressing upon the company the belief that the effect is produced by machinery, and of giving a signal to the player to put in motion the head of the automaton.

This ingenious explanation of the chess automaton is, our author states, greatly confirmed by the *regular and undeviating* mode of disclosing the interior of the chest ; and he also shows that the facts which have been observed respecting the winding up of the machine, " afford positive proof that the axis turned by the key is quite free and unconnected either with a spring or weight, or any system of machinery."

In order to make the preceding description more intelligible, I shall add the following more detailed explanation of the figures.

Fig. 70.   *Fig.* 71.

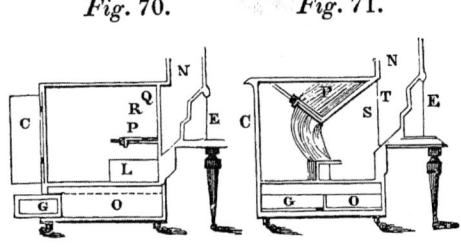

Fig. 61 is a perspective view of the automaton seen in front with all the doors thrown open.

Fig. 62 is an elevation of the automaton, as seen from behind.

Fig. 63 is an elevation of the front of the chest, the shaded figure representing the enclosed player in his first position, or when the door A is opened.

Fig. 64 is a side elevation, the shaded figure representing the player in the same position.

Fig. 65 is a front elevation, the shaded figure showing the player in his second position, or that which he takes after the door B and screen I are closed, and the great cupboard opened.

Fig. 66 is a front elevation, the shaded figure showing the player in his third position, or that in which he plays the game.

Fig. 67 is a side elevation, showing the figure in the same position.

Fig. 68 is an horizontal section of the chest through the line W W in Fig. 66.

Fig. 69 is a vertical section of the chest through the line X X in Fig. 68.

Fig. 70 is a vertical section through the line Y Y, Fig. 66 showing the false back closed.

Fig. 71 is a similar vertical section showing the false back raised.

The following letters of reference are employed in all the figures :

A. Front door of the small cupboard.

B. Back door of ditto.

C C. Front doors of large cupboard.

D. Back door of ditto.

E. Door of ditto.

F. Door of the thigh.

G G. The drawer.

H. Machinery in front of the small cupboard.

I. Screen behind the machinery.

K. Opening caused by the removal of part of the floor of the small cupboard.

L. A box which serves to conceal an opening in the floor of the large cupboard, made to facilitate the first position ; and which also serves as a seat for the third position.

M. A similar box to receive the toes of the player in the first position.

N. The inner chest filling up part of the trunk.

O. The space behind the drawer.

P Q. The false back turning on a joint at Q.

R. Part of the partition formed of cloth stretched tight, which is carried up by the false back to form the opening between the chambers.

S. The opening between the chambers.

T. The opening connecting the trunk and chest, which is partly concealed by the false back.

U. Panel which is slipt aside to admit the player.

Various pieces of mechanism of wonderful ingenuity have been constructed for the purposes of drawing and writing. One of these, invented by M. Le Droz, the son of the celebrated Droz of Chaux le Fonds, has been described by Mr Collinson. The figure was the size of life. It held in its hand a metallic style, and when a spring was touched, so as to release a detent, the figure immediately began to draw upon a card of Dutch vellum previously laid under its hand. After the drawing was executed on the first card, the figure rested. Other five cards were then put in in succession, and upon these it delineated in the same manner different subjects. On the first card it drew " elegant portraits and likenesses of the king and queen facing each other ;" and Mr Collinson remarks, that it was curious to observe with what precision the figure lifted up its pencil in its transition from one point of the drawing to another without making the slightest mistake.

M. Maillardet has executed an automaton which both writes and draws. The figure of a boy kneeling on one knee holds a pencil in his hand. When the figure begins to work, an attendant dips the pencil in ink, and adjusts the drawing-paper upon a brass tablet. Upon touching a spring, the figure proceeds to write, and when the line is finished its hand returns to dot and stroke the letters when necessary. In this manner it executes four beautiful pieces of writing in French and English, and three landscapes, all of which occupy about one hour.

One of the most popular pieces of mechanism which we have seen is the magician constructed by M. Maillardet for the purpose of answering certain given questions. A figure, dressed like a magician, appears seated at the bottom of a wall, holding a wand in one hand, and a book in the other. A number of questions ready prepared are inscribed on oval medallions, and the spectator takes any of these which he chooses, and to which he wishes an answer, and having placed it in a drawer ready to receive it, the drawer shuts with a spring till the answer is returned. The magician then rises from his seat, bows his head, describes circles with his wand, and, consulting the book as if in deep thought, he lifts it toward his face. Having thus appeared to ponder over the proposed question, he raises his wand, and striking with it the wall above his head, two folding-doors fly open, and display an appropriate answer to the question. The doors again close, the magician resumes his original position, and the drawer opens to return the medallion. There are twenty of these medallions, all containing different questions, to which the magician returns the most suitable and striking answers. The medallions are thin plates of brass of an elliptical form, exactly resembling each other. Some of the medallions have a question inscribed on each side, both of which the magician answers in succession. If the drawer is shut without a medallion being put into it, the magician rises, consults his book, shakes his head, and resumes his seat. The folding-doors remain shut, and the drawer is returned empty.

If two medallions are put into the drawer together, an answer is returned only to the lower one. When the machinery is wound up, the movements continue about an hour, during which time about fifty questions may be answered. The inventor stated, that the means by which the different medallions acted upon the machinery, so as to produce the proper answers to the questions which they contained, were extremely simple. *

The same ingenious artist has constructed various other automata representing insects and other animals. One of these was a spider entirely made of steel, which exhibited all the movements of the animal. It ran on the surface of a table during three minutes, and to prevent it from running off, its course always tended towards the centre of the table. He constructed likewise a caterpillar, a lizard, a mouse, and a serpent. The serpent crawls about in every direction, opens its mouth, hisses and darts out its tongue.

Ingenious and beautiful as all these pieces of mechanism are, and surprising as their effects appear even to scientific spectators, the principal object of their inventors was to astonish and amuse the public. We should form an erroneous judgment, however, if we supposed that this was the only result of the ingenuity which they displayed. The passion for automatic exhibitions which characterized the 18th century, gave rise to the most ingenious mechanical devices, and introduced among the higher orders of artists habits of nice and accurate execution in the formation of the most delicate pieces

* See the Edinburgh Encyclopædia, Art. *Androides,* Vol. ii. p. 66.

of machinery. The same combination of the mechanical powers which made the spider crawl, or which waved the tiny rod of the magician, contributed in future years to purposes of higher import. Those wheels and pinions, which almost eluded our senses by their minuteness, reappeared in the stupendous mechanism of our spinning-machines, and our steam-engines. The elements of the tumbling puppet were revived in the chronometer, which now conducts our navy through the ocean ; and the shapeless wheel which directed the hand of the drawing automaton has served in the present age to guide the movements of the tambouring engine. Those mechanical wonders which in one century enriched only the conjuror who used them, contributed in another to augment the wealth of the nation ; and those automatic toys which once amused the vulgar, are now employed in extending the power and promoting the civilization of our species. In whatever way, indeed, the power of genius may invent or combine, and to whatever low or even ludicrous purposes that invention or combination may be originally applied, society receives a gift which it can never lose ; and though the value of the seed may not be at once recognized, and though it may lie long unproductive in the ungenial till of human knowledge, it will some time or other evolve its germ, and yield to mankind its natural and abundant harvest.

Did the limits of so popular a volume as this ought to be, permit it, I should have proceeded to give a general description of some of these extraordinary pieces of machinery, the construction and effects of which never fail to strike the spectator with surprise. This, however, would lead me into a field too extensive, and I shall therefore confine myself to a notice of three very remarkable pieces of mechanism which are at present very little known to the general reader, viz. the tambouring machine of Mr Duncan, the statue turning machine of Mr Watt, and the calculating machinery of Mr Babbage.

The tambouring of muslins, or the art of producing upon them ornamental flowers and figures, has been long known and practised in Britain, as well as in other countries; but it was not long before the year 1790 that it became an object of general manufacture in the west of Scotland, where it was chiefly carried on. At first it was under the direction of foreigners; but their aid was not long necessary, and it speedily extended to such a degree as to occupy, either wholly or partially, more than 20,000 females. Many of these labourers lived in the neighbourhood of Glasgow, which was the chief seat of the manufacture; but others were scattered through every part of Scotland, and supplied by agents with work and money. In Glasgow, a tambourer of ordinary skill could not in general earn more than five or six shillings a week by constant application; but to a labouring artisan, who had several daughters, even these low wages formed a source of great wealth. At the age of five years, a child capable of handling a needle was devoted to tambouring, even though it could not earn more than a shilling or two in a week; and the consequence of this was, that female children were taken from school, and rendered totally unfit for any social or domestic duty. The tambouring population was therefore of the worst kind,

and it must have been regarded as a blessing rather
than as a calamity, when the work which they performed was entrusted to regular machinery.

Mr John Duncan of Glasgow, the inventor of
the tambouring machinery, was one of those unfortunate individuals who benefit their species without benefiting themselves, and who died in the meridian of life the victim of poverty and of national
ingratitude.   He conceived the idea of bringing
into action a great number of needles at the same
time, in order to shorten the process by manual
labour, but he at first was perplexed about the di-
versification of the pattern.   This difficulty, however, he soon surmounted by employing two forces
at right angles to each other, which gave him a new
force in the direction of the diagonal of the paral-
lelogram, whose sides were formed by the original
forces.   His first machine was very imperfect ; but
after two year's study he formed a company, at whose
expence six improved machines were put in action,
and who secured the invention by a patent.   At
this time the idea of rendering the machine automatic had scarcely occurred to him; but he afterwards
succeeded in accomplishing this great object, and
the tambouring machines were placed under the sur-
veillance of a steam-engine.   Another patent was
taken for these improvements.   The reader who
desires to have a minute account of these improvements, and of the various parts of the machinery,
will be amply gratified by perusing the inventor's
own account of the machinery in the article CHAIN-
WORK in the Edinburgh Encyclopædia.   At present it will be sufficient to state, that the muslin to
be tamboured was suspended vertically in a frame,

which was capable of being moved both in a vertical and a horizontal direction. Sixty or more needles lying horizontal occupied a frame in front of the muslin web. Each of these working-needles, as they are called, was attended by a feeding-needle, which by a circular motion round the working-needle lodged upon the stem of the latter the loop of the thread. The sixty needles then penetrated the web, and in order that they might return again without injuring the fabric, the barb or eye of the needle, which resembled the barb of a fishing-hook, was shut by a slider. The muslin web then took a new position by means of the machinery that gave it its horizontal and vertical motion, so that the sixty needles penetrated it, at their next movement, at another point of the figure or flower. This operation went on till sixty flowers were completed. The web was then slightly wound up, that the needles might be opposite that part of it on which they were to work another row of flowers.

The flowers were generally at an inch distance, and the rows were placed so that the flowers formed what are called diamonds. There were seventy-two rows of flowers in a yard, so that in every square yard there were nearly 4000 flowers, and in every piece of ten yards long 40,000. The number of loops or stitches in a flower varied with the pattern, but on an average there was about thirty. Hence the number of stitches in a yard were 120,000, and the number in a piece is 1,200,000. The average work done in a week by one machine was fifteen yards, or 60,000 flowers, or 1,800,000 stitches, and by comparing this with the work done by one person with the hand, it appeared that the machine

T

enabled one person to do the work of twenty-four persons.

One of the most curious and important applications of machinery to the arts which has been suggested in modern times, was made by the late Mr Watt, in the construction of a machine for copying or reducing statues and sculpture of all kinds. The art of multiplying busts and statues, by casts in plaster of Paris, has been the means of diffusing a knowledge of this branch of the fine arts ; but from the fragile nature of the material, the copies thus produced were unfit for exposure to the weather, and therefore ill calculated for ornamenting public buildings, or for perpetuating the memory of public achievements. A machine, therefore, which is capable of multiplying the labours of the sculptor in the durable materials of marble or of brass, was a desideratum of the highest value, and one which could have been expected only from a genius of the first order. During many years Mr Watt carried on his labours in secret, and he concealed even his intention of constructing such a machine. After he had made considerable progress in its execution, and had thought of securing his invention by a patent, he learned that an ingenious individual in his own neighbourhood had been long occupied in the same pursuit ; and Mr Watt informed me, that he had every reason to believe that this gentleman was entirely ignorant of his labours. A proposal was then made that the two inventors should combine their talents, and secure the privilege by a joint patent ; but Mr Watt had experienced so frequently the fatal operation of our patent laws, that he saw many difficulties in the way of such an arrangement, and he was unwilling, at his advanced age, to embark in a project so extensive, and which seemed to require for its successful prosecution all the ardour and ambition of a youthful mind. The scheme was therefore abandoned; and such is the unfortunate operation of our patent laws, that the circumstance of two individuals having made the same invention, has prevented both from bringing it to perfection, and conferring a great practical benefit upon their species. The machine which Mr Watt had constructed had actually executed some excellent pieces of work. I have seen in his house at Heathfield copies of basso relievos, and complete statues of a small size; and some of his friends have in their possession other specimens of its performance.

Of all the machines which have been constructed in modern times, the calculating-machine is doubtless the most extraordinary. Pieces of mechanism for performing particular arithmetical operations have been long ago constructed, but these bear no comparison either in ingenuity or in magnitude to the grand design conceived and nearly executed by Mr Babbage. Great as the power of mechanism is known to be, yet we venture to say that many of the most intelligent of our readers will scarcely admit it to be possible that astronomical and navigation tables can be accurately computed by machinery; that the machine can itself correct the errors which it may commit; and that the results of its calculations, when absolutely free from error, can be printed off, without the aid of human hands, or the operation of human intelligence. All this, however, Mr Babbage's machine can do; and as

I have had the advantage of seeing it actually calculate, and of studying its construction with Mr Babbage himself, I am able to make the above statement on personal observation.   The calculating-machine now constructing under the superintendence of the inventor has been executed at the expence of the British Government, and is of course their property.   It consists essentially of two parts, a calculating part, and a printing part, both of which are necessary to the fulfilment of Mr Babbage's views, for the whole advantage would be lost if the computations made by the machine were copied by human hands and transferred to types by the common process.   The greater part of the calculating-machinery is already constructed, and exhibits workmanship of such extraordinary skill and beauty that nothing approaching to it has been witnessed.   In order to execute it, particularly those parts of the apparatus which are dissimilar to any used in ordinary mechanical constructions, tools and machinery of great expense and complexity have been invented and constructed; and in many instances contrivances of singular ingenuity have been resorted to, which cannot fail to prove extensively useful in various branches of the mechanical arts.

The drawings of this machinery, which form a large part of the work, and on which all the contrivance has been bestowed, and all the alterations made, cover upwards of 400 *square feet of surface*, and are executed with extraordinary care and precision.

In so complex a piece of mechanism, in which interrupted motions are propagated simultaneously along a great variety of trains of mechanism, it might have been supposed that obstructions would arise, or even incompatibilities occur, from the impracticability of foreseeing all the possible combinations of the parts; but this doubt has been entirely removed, by the constant employment of a system of mechanical notation invented by Mr Babbage, which places distinctly in view, at every instant, the progress of motion through all the parts of this or any other machine, and by writing down in tables the times required for all the movements, this method renders it easy to avoid all risk of two opposite actions arriving at the same instant at any part of the engine.

In the printing part of the machine less progress has been made in the actual execution than in the calculating part. The cause of this is the greater difficulty of its contrivance, not for transferring the computations from the calculating part to the copper or other plate destined to receive it, but for giving to the plate itself that number and variety of movements which the forms adopted in printed tables may call for in practice.

The practical object of the calculating engine is to compute and print a great variety and extent of astronomical and navigation tables, which could not be done without enormous intellectual and manual labour, and which, even if executed by such labour, could not be calculated with the requisite accuracy. Mathematicians, astronomers, and navigators do not require to be informed of the real value of such tables; but it may be proper to state, for the information of others, that *seventeen* large folio volumes of logarithmic tables alone were calculated at an enormous expence by the French Government; and that the British Government

regarded these tables to be of such national value
that they proposed to the French Board of Lon-
gitude to print an *abridgement* of them at
the joint expence of the two nations, and offered
to advance L.5000 for that purpose. Besides logarithmic tables, Mr Babbage's machine will calculate tables of the powers and products of numbers, and
all astronomical tables for determining the positions of the sun, moon, and planets ; and the same
mechanical principles have enabled him to integrate
innumerable equations of finite differences, that is,
when the equation of differences is given, he can, by
setting an engine, produce at the end of a given
time any distant term which may be required, or
any succession of terms commencing at a distant
point.

Beside the cheapness and celerity with which
this machine will perform its work, the *absolute
accuracy* of the printed results deserves especial
notice. By peculiar contrivances, any small error
produced by accidental dust, or by any slight inac-
curacy in one of the wheels, is corrected as soon as
it is transmitted to the next, and this is done in
such a manner as effectually to prevent any ac-
cumulation of small errors from producing an erroneous figure in the result.

In order to convey some idea of this stupendous
undertaking, we may mention the effects produced
by a small trial engine constructed by the inventor,
and by which he computed the following table from
the formula $x^2 + x + 41$. The figures as they were
calculated by the machine were not exhibited to
the eye as in sliding rules and similar instruments,
but were actually presented to the eye on two op-

posite sides of the machine, the number 383, for example, appearing in figures before the person employed in copying.

Table calculated by a small Trial Engine.

41	131	383	797	1373
43	151	421	853	1447
47	173	461	911	1523
53	197	583	971	1601
61	223	547	1033	1681
71	251	593	1097	1763
83	281	641	1163	1847
97	313	691	1231	1933
113	347	743	1301	2021

While the machine was occupied in calculating this table, a friend of the inventor undertook to write down the numbers as they appeared. In consequence of the copyist writing quickly, he rather more than kept pace with the engine, but as soon as five figures appeared, the machine was at least equal in speed to the writer. At another trial *thirty-two* numbers of the same table were calculated in the space of *two minutes and thirty seconds*, and as these contained *eighty-two* figures, the engine produced thirty-three figures every minute or more than one figure in every two seconds. On another occasion it produced *forty-four* figures per minute. This rate of computation could be maintained for any length of time; and it is probable that few writers are able to copy with equal speed for many hours together.

Some of that class of individuals who envy all great men, and deny all great inventions, have ignorantly stated that Mr Babbage's invention is not new. The same persons, had it suited their purpose, would have maintained that the invention of spectacles was an anticipation of the telescope; but even this is more true than the allegation, that the arithmetical machines of Pascal and others were the types of Mr Babbage's engine. The object of these machines was entirely different. Their highest functions were to perform the operations of common arithmetic. Mr Babbage's engine, it is true, can perform these operations also, and can extract the roots of numbers, and approximate to the roots of equations, and even to their impossible roots. But this is not its object. Its function, in contradistinction to that of all other contrivances for calculating, is to embody in machinery the method of differences, which has never before been done; and the effects which it is capable of producing, and the works which in the course of a few years we expect to see it execute, will place it at an infinite distance from all other efforts of mechanical genius. *

* A popular account of this engine will be found in Mr Babbage's interesting volume *On the Economy of Manufactures*, just published.

LETTER XII.

Wonders of chemistry—Origin, progress, and objects of alche-
my—Art of breathing fire—Employed by Barchochebas, Eu-
nus, &c.—Modern method—Art of walking upon burning
coals and red hot iron, and of plunging the hands in melted
lead and boiling water—Singular property of boiling tar—
Workmen plunge their hands in melted copper—Trial of
ordeal by fire—Aldini's incombustible dresses—Examples of
their wonderful power in resisting flame—Power of breathing
and enduring air of high temperatures—Experiments made
by Sir Joseph Banks, Sir Charles Blagden, and Mr Chantry.

THE science of chemistry has from its infancy been
pre-eminently the science of wonders. In her laboratory the alchemist and the magician have re-
velled uncontrolled, and from her treasures was
forged the sceptre which was so long and so fatally
wielded over human reason. The changes which
take place in the bodies immediately around us are
too few in number and too remote from observation
to excite much of our notice; but when the substances procured directly from nature, or formed ca-
sually by art, become objects of investigation, they
exhibit in their simple or combined actions the
most extraordinary effects. The phenomena which
they display, and the products which they form, so
little resemble those with which we are familiar,

that the most phlegmatic and the least speculative
observer must have anticipated from them the creation of new and valuable compounds. It can scarcely, therefore, be a matter of surprise that minds of
the highest order, and spirits of the loftiest ambition, should have sought in the transmutations of chemistry for those splendid products which were conceived to be most conducive to human happiness.

The disciple of Mammon grew pale over his cru-
cible in his ardour to convert the baser metals into
gold :—The philosopher pined in secret for the universal solvent which might develope the elements
of the precious stones, and yield to him the means
of their production ; and the philanthropist aspired
after an universal medicine, which might arrest disease in its course, and prolong indefinitely the life
of man. To us who live under the meridian of
knowledge, such expectations must appear as presumptuous as they were delusive : but when we consider that gold and silver were actually produced by
chemical processes from the rude ores of lead and
copper ;—that some of the most refractory bodies
had yielded to the disintegrating and solvent powers
of chemical agents ;—and that the mercurial prepara-
tions of the Arabian physicians had operated like
charms in the cure of diseases that had resisted the
feeble medicines of the times, we may find some
apology for the extravagant expectations of the al-
chemists.

An object of lofty pursuit, even if it be one of
impossible attainment, is not unworthy of philosophical ambition. Though we cannot scale the summit of the volcanic cone, we may yet reach its heav-
ing flanks, and though we cannot decompose its

loftiest fires, we may yet study the lava which they have melted and the products which they have sublimed. In like manner, though the philosopher's stone has not been found, chemistry has derived rich accessions from its search ;—though the general solvent has not been obtained, yet the diamond and the gems have surrendered to science their adamantine strength ;—and though the elixir of life has never been distilled, yet other medicines have soothed " the ills which flesh is heir to," and prolonged in no slight degree the average term of our existence.

Thus far the pursuits of the alchemist were honourable and useful ; but when his calling was followed, as it soon was, by men prodigal of fortune and of character, science became an instrument of crime; secrets unattained were bartered for the gold of the credulous and the ignorant, and books innumerable were composed to teach these pretended secrets to the world. An intellectual reaction, however, soon took place, and those very princes who had sought to fill their exhausted treasuries at the furnace of the chemist, were the first to enact laws against the frauds which they had encouraged, and to dispel the illusions which had so long deceived their subjects.

But even when the moral atmosphere of Europe was thus disinfected, chemistry supplied the magician with his most lucrative wonders, and those who could no longer delude the public with dreams of wealth and longevity, now sought to amuse and astonish them by the exhibition of their skill. The narrow limits of this volume will not permit me to give even a general view of those extraordinary effects which this popular science can display.

I must therefore select from its inexhaustible stores those topics which are most striking in their results, and most popular in their details.

One of the most ancient feats of magic was the art of breathing flame,—an art which even now excites the astonishment of the vulgar. During the insurrection of the slaves in Sicily in the second century before Christ, a Syrian named Eunus acquired by his knowledge the rank of their leader. In order to establish his influence over their minds, he pretended to possess miraculous power. When he wished to inspire his followers with courage, he breathed flames or sparks among them from his mouth, at the same time that he was rousing them by his eloquence. St Jerome informs us, that the Rabbi Barchochebas, who headed the Jews in their last revolt against Hadrian, made them believe that he was the Messiah, by vomiting flames from his mouth ; and at a later period, the Emperor Constantius was thrown into a state of alarm when Valentinian informed him, that he had seen one of the body guards breathing out fire and flames. We are not acquainted with the exact methods by which these effects were produced ; but Florus informs us, that Eunus filled a perforated nut-shell with sulphur and fire, and having concealed it in his mouth, he breathed gently through it while he was speaking. This art is performed more simply by the modern juggler. Having rolled together some flax or hemp, so as to form a ball the size of a walnut, he sets it on fire, and allows it to burn till it is nearly consumed : He then rolls round it while burning some additional flax, and by these means the fire may be retained in it for a considerable time. At the commencement of his exhibition he introduces the ball into his mouth, and while he breathes through it the fire is revived, and a number of burning sparks are projected from his mouth. These sparks are too feeble to do any harm, provided he inhales the air through his nostrils.

The kindred art of walking on burning coals or red hot iron remounts to the same antiquity. The priestesses of Diana at Castabala in Cappadocia were accustomed, according to Strabo, to walk over burning coals; and at the annual festival, which was held in the temple of Apollo on Mount Soracte in Etruria, the Hirpi marched over burning coals, and on this account they were exempted from military service, and received other privileges from the Roman Senate. This power of resisting fire was ascribed even by Varro to the use of some liniment with which they anointed the soles of their feet.

Of the same character was the art of holding red hot iron in the hands or between the teeth, and of plunging the hands into boiling water or melted lead. About the close of the seventeenth century, an Englishman of the name of Richardson rendered himself famous by chewing burning coals, pouring melted lead upon his tongue, and swallowing melted glass. That these effects are produced partly by deception, and partly by a previous preparation of the parts subjected to the heat, can scarcely admit of a doubt. The fusible metal, composed of mercury, tin, and bismuth, which melts at a low temperature, might easily have been substituted in place of lead; and fluids of easy ebullition may have been used in place of boiling water. A solution of spermaceti or sulphuric ether, tinged with alkanet root, which becomes solid at 50° of Fahrenheit, and melts and boils with the heat of the hand, is supposed to be the substance which is used at Naples when the dried blood of St Januarius melts spontaneously, and boils over the vessel which contains it.

But even when the fluid requires a high temperature to boil, it may have other properties, which enable us to plunge our hands into it with impunity. This is the case with boiling tar, which boils at a temperature of 220° even higher than that of water. Mr Davenport informs us, that he saw one of the workmen in the King's Dock-yard at Chatham immerse his naked hand in tar of that temperature. He drew up his coat sleeves, dipped in his hand and wrist, bringing out fluid tar, and pouring it off from his hand as from a ladle. The tar remained in complete contact with his skin, and he wiped it off with tow. Convinced that there was no deception in this experiment, Mr Davenport immersed the entire length of his fore-finger in the boiling cauldron, and moved it about a short time before the heat became inconvenient. Mr Davenport ascribes this singular effect to the slowness with which the tar communicates its heat, which he conceives to arise from the abundant volatile vapour which is evolved " carrying off rapidly the caloric in a latent state, and intervening between the tar and the skin, so as to prevent the more rapid communication of heat." He conceives also, that when the hand is withdrawn, and the hot tar adhering to it, the rapidity with which this vapour is evolved from the surface exposed to the air cools it immediately. The workmen informed Mr Davenport, that, if a person put his hand into the cauldron with his glove on, he would be dreadfully burnt, but this extraordinary result was not put to the test of observation.

But though the conjurors with fire may have availed themselves of these singular properties of individual bodies, yet the general secret of their art consisted in rendering the skin of the exposed parts callous and insensible to heat,—an effect which may be produced by continually compressing or singeing them till the skin acquires a horny consistence. A proof of this opinion is mentioned by Beckmann, who assures us that in September 1765, when he visited the copper works at Awestad, one of the workmen, bribed by a little money to drink, took some of the melted copper in his hand, and after shewing it to the company, threw it against a wall. He then squeezed the fingers of his horny hand close to each other, held it a few minutes under his arm-pit to make it perspire, as he said, and taking it again out, drew it over a ladle filled with melted copper, some of which he skimmed off, and moved his hand backwards and forwards very quickly by way of ostentation. During this performance, M. Beckmann noticed a smell like that of singed horn or leather, though the hand of the workman was not burned. This callosity of the skin may be effected by frequently moistening it with dilute sulphuric acid. Some allege that the juices of certain plants produce the same effect, while others recommend the frequent rubbing of the skin with oil. The receipt given by Albertus Magnus for this purpose was of a different nature. It consisted of a non-conducting calcareous paste, which was made to adhere to the skin by the sap of the marsh-mallow, the slimy seeds of the flea-bane, and the white of an egg.

As the ancients were acquainted with the incombustibility of asbestos or amianthus, and the art of weaving it into cloth, it is highly probable that it was employed in the performance of some of their miracles, and it is equally probable that it was subsequently used, along with some of the processes already described, in enabling the victims of superstition to undergo without hazard the trial of ordeal by fire. In every country where this barbarous usage prevailed, whether in the sanctuary of the Christian idolater, or in the pagan temple of the Bramin, or under the wild orgies of the African savage, Providence seems to have provided the means of meeting it with impunity. In Catholic countries this exculpatory judgment was granted chiefly to persons in weak health, who were incapable of using arms, and particularly to monks and ecclesiastics who could not avail themselves of the trial by single combat. The fire ordeal was conducted in the church under the inspection of the clergy: Mass was at the same time celebrated, and the iron and the victims were consecrated by the sprinkling of holy water. The preparatory steps were also under the direction of the priests. It was necessary that the accused should be placed three days and three nights under their care, both before and after the trial. Under the pretence of preventing the defendant from preparing his hands by art; and in order to ascertain the result of the ordeal, his hands were covered up and sealed during the three days which preceded and followed the fiery application; and it has been plausibly conjectured by Beckmann, that during the three first days the preventative was applied to those whom they wished to acquit, and that the three last days were requisite to bring back the hands to their natural condition. In these and other cases, the accused could not have availed himself directly of the use of asbestos gloves, unless we could suppose them so made as to imitate the human skin at a distance; but the fibres of that mineral may have been imbedded in a paste which applied itself readily to all the elevations and depressions of the skin.

In our own times the art of defending the hands and face, and indeed the whole body, from the action of heated iron and intense fire, has been applied to the nobler purpose of saving human life, and rescuing property from the flames. The revival and the improvement of this art we owe to the benevolence and the ingenuity of the Chevalier Aldini of Milan, who has travelled through all Europe to present this valuable gift to his species. Sir H. Davy had long ago shewn that a safety lamp for illuminating mines, containing inflammable air, might be constructed of wire-gauze alone, which prevented the flame within, however large or intense, from setting fire to the inflammable air without. This valuable property, which has been long in practical use, he ascribed to the conducting and radiating power of the wire-gauze, which carried off the heat of the flame, and deprived it of its power. The Chevalier Aldini conceived the idea of applying the same material in combination with other badly conducting substances, as a protection against fire. The incombustible pieces of dress which he uses for the body, arms,

and legs, are formed out of strong cloth, which has
been steeped in a solution of alum, while those for
the head, hands, and feet, are made of cloth of asbestos or amianthus.   The head dress is a large cap
which envelopes the whole head down to the neck,
having suitable perforations for the eyes, nose, and
mouth.   The stockings and cap are single, but the
gloves are made of double amianthus cloth, to enable the fireman to take into his hand burning or
red hot bodies.   The piece of ancient asbestos cloth
preserved in the Vatican was formed, we believe, by
mixing the asbestos with other fibrous substances ;
but M. Aldini has executed a piece of nearly the
same size, nine feet five inches long, and five feet
three inches wide, which is much stronger than the
ancient piece, and possesses superior qualities, in
consequence of having been woven without the introduction of any foreign substance.   In this manufacture the fibres are prevented from breaking by
the action of steam, the cloth is made loose in its
fabric, and the threads are about the fiftieth of an
inch in diameter.

The metallic dress which is superadded to these
means of defence consists of five principal pieces,
viz. a *casque* or cap, with a mask large enough to
leave a proper space between it and the asbestos
cap ; a cuirass with its brassets ; a piece of armour
for the trunk and thighs ; a pair of boots of double
wire-gauze ; and an oval shield five feet long by $2\frac{1}{2}$
wide, made by stretching the wire-gauze over a
slender frame of iron.   All these pieces are made of
iron wire-gauze, having the interval between its
threads the twenty-fifth part of an inch.

In order to prove the efficacy of this apparatus,

and inspire the firemen with confidence in its protection, he showed them that a finger first enveloped in asbestos, and then in a double case of wire-
gauze, might be held a long time in the flame of a
spirit-lamp or candle before the heat became inconvenient. A fireman having his hand within a
double asbestos glove, and its palm protected by a
piece of asbestos cloth, seized with impunity a large
piece of red hot iron, carried it deliberately to the
distance of 150 feet, inflamed straw with it, and
brought it back again to the furnace.   On other
occasions, the firemen handled blazing wood and
burning substances, and walked during five minutes
upon an iron grating placed over flaming faggots.

   In order to show how the head, eyes, and lungs,
are protected, the fireman put on the asbestos and
wire-gauze cap, and the curiass, and held the shield
before his breast. A fire of shavings was then
lighted, and kept burning in a large raised chafing-
dish, and the fireman plunged his head into the
middle of the flames with his face to the fuel, and
in that position went several times round the
chafing-dish for a period longer than a minute.   In
a subsequent trial at Paris, a fireman placed his
head in the middle of a large brasier filled with
flaming hay and wood, as in Fig. 72, and resisted
the action of the fire during five or six minutes,
and even ten minutes.

   In the experiments which were made at Paris
in presence of a committee of the Academy of
Sciences, two parallel rows of straw and brushwood,
supported by iron wires, were formed at the distance of three feet from each other, and extended
thirty feet in length.   When this combustible mass

was set on fire, it was necessary to stand at the distance of eight or ten yards to avoid the heat.   The

Fig. 72.

flames from both the rows seemed to fill up the
whole space between them, and rose to the height
of nine or ten feet.   At this moment six firemen,
clothed in the incombustible dresses, and marching
at a slow pace behind each other, repeatedly passed
through the whole length between the two rows
of flame, which were constantly fed with additional
combustibles.   One of the firemen carried on his
back a child eight years old in a wicker basket covered with metallic gauze, and the child had no
other dress than a cap made of amianthine cloth.
    In February 1829, a still more striking experi-

4

ment was made in the yard of the barracks of St Gervais. Two towers were erected two stories high, and were surrounded with heaps of inflamed materials, consisting of faggots and straw. The firemen braved the danger with impunity. In opposition to the advice of M. Aldini, one of them, with the basket and child, rushed into a narrow place, where the flames were raging eight yards high. The violence of the fire was so great that he could not be seen, while a thick black smoke spread around, throwing out a heat which was unsupportable by the spectators. The fireman remained so long invisible that serious doubts were entertained of his safety. He at length, however, issued from the fiery gulf uninjured, and proud of having succeeded in braving so great a danger.

It is a remarkable result of these experiments, that the firemen are able to breathe without difficulty in the middle of the flames. This effect is owing not only to the heat being intercepted by the wire gauze as it passes to the lungs, in consequence of which its temperature becomes supportable, but also to the singular power which the body possesses of resisting great heats, and of breathing air of high temperatures.

A series of curious experiments were made on this subject by M. Tillet, in France, and by Dr Fordyce and Sir Charles Blagden, in England. Sir Joseph Banks, Dr Solander, and Sir Charles Blagden, entered a room in which the air had a temperature of 198° Fahr. and remained ten minutes ; but as the thermometer sunk very rapidly, they resolved to enter the room singly. Dr Solander went in alone, and found the heat 210°, and

Sir Joseph entered when the heat was 211°.
Though exposed to such an elevated temperature,
their bodies preserved their natural degree of heat.
Whenever they breathed upon a thermometer it
sunk several degrees : Every expiration, particularly if strongly made, gave a pleasant impression
of coolness to their nostrils, and their cold breath
cooled their fingers whenever it reached them. On
touching his side, Sir Charles Blagden found it
cold like a corpse, and yet the heat of his body under his tongue was 98°. Hence they concluded
that the human body possesses the power of de-
stroying a certain degree of heat when communicated with a certain degree of quickness. This
power, however, varies greatly in different media.
The same person who experienced no inconvenience
from air heated to 211°, could just bear rectified
spirits of wine at 130°, cooling oil at 129°, cooling
water at 123°, and cooling quicksilver at 117°. A
familiar instance of this occurred in the heated
room. All the pieces of metal there, even their
watch-chains, felt so hot, that they could scarcely
bear to touch them for a moment, while the air
from which the metal had derived all its heat was
only unpleasant. MM. Duhamel and Tillet observed at Rochefoucault in France, that the girls
who were accustomed to attend ovens in a bakehouse
were capable of enduring for ten minutes a temperature of 270°.

The same gentlemen who performed the experiments above described ventured to expose themselves to still higher temperatures. Sir Charles
Blagden went into a room where the heat was 1°
or 2° above 260°, and remained eight minutes in

this situation, frequently walking about to all the different parts of the room, but standing still most of the time in the coolest spot, where the heat was above 240°. The air, though very hot, gave no pain, and Sir Charles and all the other gentlemen were of opinion that they could support a much greater heat. During seven minutes, Sir C. Blagden's breathing continued perfectly good, but after that time he felt an oppression in his lungs, with a sense of anxiety, which induced him to leave the room. His pulse was then 144, double its ordinary quickness. In order to prove that there was no mistake respecting the degree of heat indicated by the thermometer, and that the air which they breathed was capable of producing all the well known effects of such a heat on inanimate matter, they placed some eggs and a beef-steak upon a tin frame near the thermometer, but more distant from the furnace than from the wall of the room. In the space of twenty minutes the eggs were roasted quite hard, and in forty-seven minutes the steak was not only dressed, but almost dry. Another beef-steak, similarly placed, was rather overdone in thirty-three minutes. In the evening, when the heat was still more elevated, a third beef-steak was laid in the same place, and as they had noticed that the effect of the hot air was greatly increased by putting it in motion, they blew upon the steak with a pair of bellows, and thus hastened the dressing of it to such a degree, that the greatest portion of it was found to be pretty well done in thirteen minutes.

Our distinguished countryman, Mr Chantry, has very recently exposed himself to a temperature still higher than any which we have mentioned.   The furnace which he employs for drying his moulds is about 14 feet long, 12 feet high, and 12 feet broad. When it is raised to its highest temperature, with the doors closed, the thermometer stands at 350°, and the iron floor is red hot.   The workmen often enter it at a temperature of 340°, walking over the iron floor with wooden clogs, which are of course charred on the surface.   On one occasion Mr Chantry, accompanied by five or six of his friends, entered the furnace, and, after remaining two minutes, they brought out a thermometer which stood at 320°.   Some of the party experienced sharp pains in the tips of their ears, and in the septum of the nose, while others felt a pain in their eyes.

LETTER XIII.

Spontaneous combustion—In the absorption of air by powder-ed charcoal—and of hydrogen by spongy platinum—Doberei-ner's lamp—Spontaneous combustion in the bowels of the earth—Burning cliffs—Burning soil—Combustion without flame—Spontaneous combustion of human beings—Countess Zangari—Grace Pett—Natural fire temples of the Guebres —Spontaneous fires in the Caspian Sea—Springs of inflam-mable gas near Glasgow—Natural light-house of Maracay-bo—New elastic fluids in the cavities of gems—Chemical operation going on in their cavities—Explosions produced in them by heat—Remarkable changes of colour from che-mical causes—Effects of the nitrous oxide or paradise gas when breathed—Remarkable cases described.—Conclusion.

AMONG the wonderful phenomena which chemistry presents to us, there are few more remarkable than those of spontaneous combustion, in which bodies, both animate and inanimate, emit flames, and are sometimes entirely consumed by internal fire. One of the commonest experiments in chemistry is that of producing inflammation by mixing two fluids perfectly cold. Becker, we believe, was the first person who discovered that this singular effect was produced by mixing oil of vitriol with oil of tur-pentine. Borrichios shewed that aquafortis pro-duced the same effect as oil of vitriol. Tournefort proved that spirit of nitre and oil of sassafras took

fire when mixed ; and Homberg discovered that the
same property was possessed by many volatile oils
when mixed with spirit of nitre.

Every person is familiar with the phenomena of
heat and combustion produced by fermentation.
Ricks of hay and stacks of corn have been frequently consumed by the heat generated during the
fermentation produced from moisture ; and gunpowder magazines, barns, and paper-mills have been
often burnt by the fermentation of the materials
which they contained. Galen informs us that the
dung of a pigeon is sufficient to set fire to a house,
and he assures us that he has often seen it take fire
when it had become rotten. Casati likewise relates on good authority, that the fire which consumed
the great church of Pisa was occasioned by the
dung of pigeons that had for centuries built their
nests under its roof.

Among the substances subject to spontaneous
combustion, pulverised or finely powdered charcoal is
one of the most remarkable. During the last thirty
years no fewer than four cases of the spontaneous
inflammation of powdered charcoal have taken
place in France. When charcoal is triturated in
tuns with bronze bruisers it is reduced into the state
of the finest powder. In this condition it has the
appearance of an unctuous fluid, and it occupies a
space three times less than it does in rods of about
six inches long. In this state of extreme division
it absorbs air much more readily than it does when
in rods. This absorption, which is so slow as to
require several days for its completion, is accompanied with a disengagement of heat, which rises
from 340° to 360° nearly of Fahrenheit, and which

is the true cause of the spontaneous inflammation·
The inflammation commences near the centre of
the mass, at the depth of five or six inches beneath
its surface, and at this spot the temperature is always higher than at any other.   Black charcoal
strongly distilled, heats and inflames more easily
than the orange, or that which is little distilled, or
than the charcoal made in boilers.   The most inflammable charcoal must have a mass of at least
66 lbs. avoirdupois, in order that it may be susceptible of spontaneous inflammation.   With the other
less inflammable varieties, the inflammation takes
place only in larger masses.

The inflammation of powdered charcoal is more
active in proportion to the shortness of the interval
between its carbonization and trituration.   The
free admission of air to the surface of the charcoal
is also indispensable to its spontaneous combustion.

Colonel Aubert, to whom we owe these interesting results, likewise found that when sulphur and
saltpetre are added to the charcoal, it loses its power
of inflaming spontaneously.   But as there is still
an absorption of air and a generation of heat, he is
of opinion that it would not be prudent to leave
these mixtures in too large masses after trituration. *

A species of spontaneous combustion, perfectly
analogous to that now described, but produced almost
instantaneously, was discovered by Professor Dobereiner of Jena in 1824.   He found that when a jet
of hydrogen gas was thrown upon recently prepared spongy platinum, the metal became almost in-

* See Edinburgh Journal of Science, New Series, No, viii.
p. 274.

stantly red hot, and set fire to the gas. In this case the minutely divided platinum acted upon the hydrogen gas, in the same manner as the minutely divided charcoal acted upon common air. Heat and combustion were produced by the absorption of both gases, though in the one case the effect was instantaneous, and in the other was the result of a prolonged absorption.

This beautiful property of spongy platinum was happily applied to the construction of lamps for producing an instantaneous light.   The form given to the lamp by Mr Garden of London is shown in the annexed figure, where A B is a globe of

Fig. 73.

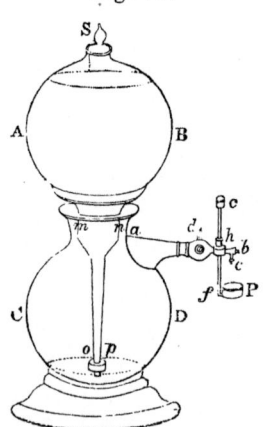

glass, fitting tightly into another glass globe C D by a ground shoulder *m n*. The globe A B terminates in a hollow tapering neck *m n o p*, on the lower end of which is placed a small cylinder of zinc *o p*. A brass tube *a b c*, is fitted at *a* into the neck of the globe C D, and through this tube, which is furnished with a stop-cock *d*, the gas can escape at the small aperture *c*. A brass pin *c f*, carrying a brass box P, is made to slide through a hole *h*, so that the brass box P, in which the spongy platinum is placed, can be set at any required distance from the aperture *c*. If sulphuric acid diluted with an equal quantity of water is now poured into the vessel A B by its mouth at S, now closed with a stopper, the fluid will descend through the tube *m n o p*, and if the cock *d* is shut, it will compress the air contained in C D. The dilute acid thus introduced into C D will act upon the ring of zinc *o p*, and generate hydrogen gas, which after the atmospheric air in C D is let off, will gradually fill the vessel C D, the diluted acid being forced up the tube *o p m n*, into the glass globe A B. The ring of zinc *o p* floats on a piece of cork, so that when C D is full of hydrogen, the diluted acid does not touch the zinc, and consequently is prevented from producing any more gas. The instant, however, that any gas is let off at *c*, the pressure of the fluid in the globe A B, and tube *m n o p*, overcomes the elasticity of the remaining gas in C D, and forces the diluted acid up to the zinc *o p*, so as to enable it to produce more gas to supply what has been used.

The lamp being supplied with hydrogen in the manner now described, it is used in the following manner: The spongy platinum in P being brought near *c*, the cock *d* is turned, and the gas is thrown

upon the platinum.   An intense heat is immediately produced, the platinum becomes red hot, and the
hydrogen enflames.   A taper is then lighted at the
flame, and the cock d is shut.   Professor Cumming
of Cambridge found it necessary to cover up the
platinum with a cap after every experiment.   This
ingenious chemist likewise found, that, with platinum
foil the 9000dth part of an inch thick kept in a
close tube, the hydrogen was inflamed; but when
the foil was only the 6000dth of an inch thick, it
was necessary to raise it previously to a red heat.

Spontaneous combustion is a phenomenon which
occurs very frequently and often to a great extent
within the bowels of the earth.   The heat by which
it is occasioned is produced by the decomposition
of mineral bodies, and other causes.   This heat increases in intensity till it is capable of melting the
solid materials which are exposed to it.   Gases and
aqueous vapours of powerful elasticity are generated, new fluids of expansive energy imprisoned in
cavities under great pressure are set free, and these
tremendous agents, acting under the repressing
forces of the superincumbent strata, exhibit their
power in desolating earthquakes; or, forcing their
way through the superficial crust of the globe, they
waste their fury in volcanic eruptions.

When the phenomena of spontaneous combustion take place near the surface of the earth its effects are of a less dangerous character, though they
frequently give birth to permanent conflagrations,
which no power can extinguish.   An example of
this milder species of spontaneous combustion has
been recently exhibited in the burning cliff at Wey-
mouth; and a still more interesting one exists at

this moment near the village of Bradley, in Staffordshire. The earth is here on fire, and this fire has continued for nearly sixty years, and has resisted every attempt that has been made to extinguish it. This fire, which has reduced many acres of land to a mere calx, arises from a burning stratum of coal about four feet thick and eight or ten yards deep, to which the air has free access, in consequence of the main coal having been dug from beneath it. The surface of the ground is sometimes covered for many yards with such quantities of sulphur that it can be easily gathered. The calx has been found to be an excellent material for the roads, and the workmen who collect it often find large beds of alum of an excellent quality.

A singular species of invisible combustion, or of combustion without flame, has been frequently noticed. I have observed this phenomenon in the small green wax tapers in common use. When the flame is blown out, the wick will continue red hot for many hours, and if the taper were regularly and carefully uncoiled, and the room kept free from currents of air, the wick would burn on in this way till the whole of the taper is consumed. The same effects are not produced when the colour of the wax is red. In this experiment the wick, after the flame is blown out, has sufficient heat to convert the wax into vapour, and this vapour, being consumed without flame, keeps the wick at its redheat. A very disagreeable vapour is produced during this imperfect combustion of the wax.

Prof. Dobereiner of Jena observed, that when the alcohol in a spirit of wine lamp was nearly exhausted, the wick became carbonized, and though the flame disappeared, the carbonized part of the

wick became red hot, and continued so while a drop
of alcohol remained, and provided the air in the
room was undisturbed.   On one occasion the wick
continued red hot for twenty-four hours, and a very
disagreeable acid vapour was formed.

On these principles depend the *lamp without
flame* which was originally constructed by Mr Ellis.
It is shown in the annexed figure, where A B is

Fig. 74.

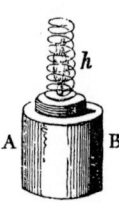

the lamp, and *h* a cylindrical coil of platinum wire,
the hundredth part of an inch in diameter.   This
spiral is so placed that four or five of the twelve
coils of which the cylinder consists are upon the
wick, and the other seven or eight above it.   If the
lamp is lighted, and continues burning till the cylindrical coil is red hot, then if the flame is blown
out, the vapour which arises from the alcohol will
by its combustion keep the coils above the wick
red hot, and this red heat will in its turn keep up
the vaporisation of the alcohol till the whole of the
alcohol is consumed.   The heat of the wire is always sufficient to kindle a piece of German fungus
or saltpetre paper, so that a sulphur match may at
any time be lighted. Mr Gill found that a wick composed of twelve threads of the cotton yarn common-

ly used for lamps will require half an ounce of alcohol to keep the wire red hot for eight hours. This lamp has been kept burning for sixty hours ; but it can scarcely be recommended for a bed-room, as an acid vapour is disengaged during the burning of the alcohol. When perfumes are dissolved in the alcohol, they are diffused through the apartment during the slow combustion of the vapour.

A species of combustion without flame, and analogous to that which has been described, is exhibited in the extraordinary phenomenon of the spontaneous combustion of living bodies. That animal bodies are liable to internal combustion is a fact which was well known to the ancients. Many cases which have been adduced as examples of spontaneous combustion are merely cases of individuals who were highly susceptible of strong electrical excitation. In one of these cases, however, Peter Bovisteau asserts, that the sparks of fire thus produced reduced to ashes the hair of a young man ; and John De Viana informs us, that the wife of Dr Freilas, physician to the Cardinal de Royas, Archbishop of Toledo, emitted by perspiration an inflammable matter of such a nature, that when the ribbon which she wore over her shift was taken from her, and exposed to the cold air, it instantly took fire, and shot forth like grains of gunpowder. Peter Borelli has recorded a fact of the very same kind respecting a peasant whose linen took fire, whether it was laid up in a box when wet, or hung up in the open air. The same author speaks of a woman who, when at the point of death, vomited flames ; and Thomas Bartholin mentions this phenomenon as having often happened to persons who were

x

great drinkers of wine or brandy. Ezekiel de Castro mentions the singular case of Alexandrinus Megetius, a physician, from one of whose vertebræ there issued a fire which scorched the eyes of the beholders ; and Krantzius relates, that during the wars of Godfrey of Boulogne, certain people of the territory of Nivers were burning with invisible fire, and that some of them cut off a foot or a hand where the burning began, in order to arrest the calamity. Nor have these effects been confined to man. In the time of the Roman consuls Gracchus and Juventius, a flame is said to have issued from the mouth of a bull without doing any injury to the animal.

The reader will judge of the degree of credit which may belong to these narrations when he examines the effects of a similar kind which have taken place in less fabulous ages, and nearer our own times. John Henry Cohausen informs us, that a Polish gentleman in the time of the Queen Bona Sforza, having drunk two dishes of a liquor called brandy-wine, vomited flames, and was burned by them, and Thomas Bartholin * thus describes a similar accident : " A poor woman at Paris used to drink spirit of wine plentifully for the space of three years, so as to take nothing else. Her body contracted such a combustible disposition, that one night, when she lay down on a straw couch, she was all burned to ashes except her skull and the extremities of her fingers." John Christ. Sturmius informs us in the German Ephemerides, that in the northern countries of Europe flames often evaporate from the stomachs of those who are addict-

* Acta Medica et Philosophica Hafniensia, 1673.

ed to the drinking of strong liquors ; and he adds,
" that seventeen years before, three noblemen of
Courland drank by emulation strong liquors, and
two of them died scorched and suffocated by a flame
which issued from their stomach."

One of the most remarkable cases of spontaneous combustion is that of the Countess Cornelia
Zangari and Bandi of Cesena, which has been minutely described by the Reverend Joseph Bianchi-
ni, a prebend in the city of Verona.   This lady,
who was in the sixty-second year of her age, retired to bed in her usual health. Here she spent above
three hours in familiar conversation with her maid
and in saying her prayers ; and having at last fallen asleep, the door of her chamber was shut.   As
her maid was not summoned at the usual hour, she
went into the bed-room to wake her mistress ; but
receiving no answer she opened the window, and
saw her corpse on the floor in the most dreadful
condition.   At the distance of four feet from the
bed there was a heap of ashes. Her legs, with
the stockings on, remained untouched, and the
head, half-burned, lay between them. Nearly all
the rest of the body was reduced to ashes. The
air in the room was charged with floating soot. A
small oil lamp on the floor was covered with ashes
but had no oil in it ; and in two candlesticks, which
stood upright upon a table, the cotton wick of both
the candles was left, and the tallow of both had
disappeared.   The bed was not injured, and the
blankets and sheets were raised on one side as if a
person had risen up from it. From an examination
of all the circumstances of this case, it has been generally supposed that an internal combustion had

taken place; that the lady had risen from her bed
to cool herself, and that, in her way to open the
window, the combustion had overpowered her, and
consumed her body by a process in which no flame
was produced which could set fire to the furniture
or the floor. The Marquis Scipio Maffei was informed by an Italian nobleman who passed through
Cesena a few days after this event, that he heard
it stated in that town, that the Countess Zangari
was in the habit, when she felt herself indisposed,
of washing all her body with camphorated spirit of
wine.

So recently as 1744 a similar example of spontaneous combustion occurred in our own country
at Ipswich. A fisherman's wife of the name of
Grace Pett, of the parish of St Clements, had been
in the habit for several years of going down stairs
every night after she was half-undressed to smoke
a pipe. She did this on the evening of the 9th
of April 1744. Her daughter, who lay in the same
bed with her, had fallen asleep, and did not miss her
mother till she awaked early in the morning. Upon
dressing herself, and going down stairs, she found
her mother's body lying on the right side with her
head against the grate, and extended over the hearth
with her legs on the deal floor, and appearing like
a block of wood burning with a glowing fire without flame. Upon quenching the fire with two
bowls of water, the neighbours, whom the cries of
the daughter had brought in, were almost stifled
with the smell. The trunk of the unfortunate
woman was almost burned to ashes, and appeared
like a heap of charcoal covered with white ashes.
The head, arms, legs, and thighs, were also much

burned.   There was no fire whatever in the grate,
and the candle was burned out in the socket of the
candlestick, which stood by her.   The clothes of a
child on one side of her, and a paper screen on the
other, were untouched ; and the deal floor was
neither singed nor discoloured.   It was said that
the woman had drunk plentifully of gin overnight
in welcoming a daughter who had recently returned from Gibraltar.

Among the phenomena of the natural world which
are related to those of spontaneous combustion, are
what have been called the natural fire temples of
the Guebres, and the igneous phenomena which
are seen in their vicinity.   The ancient sect of the
Guebres or Parsees, distinguished from all other sects
as the worshippers of fire, had their origin in Persia ; but, being scattered by persecution, they sought
an asylum on the shores of India.   Those who refused to expatriate themselves continued to inhabit
the shores of the Caspian Sea, and the cities of
Ispahan, Yezd, and Kerman.   Their great fire
temple called Attush Kudda stands in the vicinity
of Badku, one of the largest and most commodious
ports in the Caspian.   In the neighbourhood of
this town the earth is impregnated with naphtha,
an inflammable mineral oil, and the inhabitants
have no other fuel, and no other light, but what is
derived from this substance.

The remains of the ancient fire temples of the
Guebres are still visible about ten miles to the north-
east of the town.   The temple in which the Deity
is worshipped under the form of fire is a space
about thirty yards square, surrounded with a low
wall, and containing many apartments.   In each of

these a small volcano of sulphureous fire issues
from the ground through a furnace or funnel in
the shape of a Hindoo altar.   On closing the funnel the fire is instantly extinguished, and by placing the ear at the aperture a hollow sound is heard,
accompanied with a strong current of cold air, which
may be lighted at pleasure by holding to it any
burning substance.   The flame is of a pale clear
colour, without any perceptible smoke, and emits a
highly sulphureous vapour, which impedes respiration, unless when the mouth is kept beneath the
level of the furnace.   This action on the lungs
gives the Guebres a wan and emaciated appearance,
and oppresses them with a hectic cough, which
strangers also feel while breathing this insalubrious
atmosphere.

For about two miles in circumference, round the
principal fire, the whole ground, when scraped to the
depth of two or three inches, has the singular property of being inflamed by a burning coal.   In this
case, however, it does not communicate fire to the
adjacent ground : But if the earth is dug up with
a spade, and a torch brought near it, an extensive,
but instantaneous conflagration takes place, in which
houses have often been destroyed, and the lives of
the people exposed to imminent danger.

When the sky is clear and the weather serene,
the springs in their ebullition do not rise higher
than two or three feet ; but in gloomy weather,
and during the prevalence of stormy clouds, the
springs are in a state of the greatest ebullition, and
the naphtha, which often takes fire spontaneously at
the earth's surface, flows burning in great quantities
to the sea, which is frequently covered with it, in a

state of flame, to the distance of several leagues from the shore.

Besides the fires in the temple, there is a large one which springs from a natural cliff in an open situation, and which continually burns. The general space in which this volcanic fire is most abundant is somewhat less than a mile in circuit. It forms a low flat hill sloping to the sea, the soil of which is a sandy earth mixed with stones. Mr Forster did not observe any violent eruption of flame in the country around the Attush Kudda; but Kinneir informs us, that the whole country round Badku has at times the appearance of being enveloped in flames. "It often seems," he adds, "as if the fire rolled down from the mountains in large masses, and with incredible velocity; and during the clear moonshine nights of November and December, a bright blue light is observed at times to cover the whole western range. The fire does not consume, and if a person finds himself in the middle of it no warmth is felt."

The inhabitants apply these natural fires to domestic purposes, by sinking a hollow cane or merely a tube of paper, about two inches in the ground, and by blowing upon a burning coal held near the orifice of the tube, there issues a slight flame, which neither burns the cane nor the paper. By means of these canes or paper tubes, from which the fire issues, the inhabitants boil the water in their coffee urns, and even cook different articles of food. The flame is put out by merely plugging up the orifice. The same tubes are employed for illuminating houses that are not paved. The smell of naphtha is of course diffused through the house, but

after any person is accustomed to it, it ceases to be
disagreeable. The inhabitants also employ this natural fire in calcining lime. The quantity of naphtha
procured in the plain to the south east of Badku is
enormous. It is drawn from wells, some of which
yield from 1000 to 1500 ℔s per day. As soon as
these wells are emptied, they fill again till the naphtha rises to its original level. *

Inflammable gases issuing from the earth have
been used both in the old and the new world for domestic purposes. In the salt mine of Gottesgabe at
Rheims, in the county of Fecklenburg, there is a pit
called the *Pit of the Wind*, from which a constant
current of inflammable gas has issued for sixty years.
M. Roeder, the inspector of the mines, has used this
gas for two years not only as a light but for all the
purposes of domestic economy. In the pits which
are not worked, he collects the gas and conveys it in
tubes to his house. It burns with a white and brilliant flame, has a density of about 0.66, and contains
traces of carbonic acid gas and sulphuretted hydrogen. †

Near the village of Fredonia in North America,
on the shores of Lake Erie, are a number of burning springs as they are called. The inflammable gas
which issues from these springs is conveyed in pipes
to the village, which is actually lighted by them. ‡

In the year 1828, a copious spring of inflammable
gas was discovered in Scotland in the bed of a rivulet which crosses the north road between Glasgow
and Edinburgh, a little to the east of the seventh

* See Forster's Travels, and Kinneir's Geog. Memoir.
† Edinburgh Journal of Science, No. xv. p. 183.
‡ Id. Id.

mile-stone from Glasgow, and only a few hundred yards from the house of Bedlay. The gas is said to issue for more than half a mile along the banks of the rivulet. Dr Thomson, who has analysed the gas, saw it issuing only within a space about fifty yards in length, and about half as much in breadth. "The emission of gas was visible in a good many places along the declivity to the rivulet in the immediate neighbourhood of a small farm-house. The farmer had set the gas on fire in one place about a yard square, out of which a great many small jets were issuing. It had burnt without interruption during five weeks, and the soil (which was clay) had assumed the appearance of pounded brick all around.

"The flame was yellow and strong, and resembled perfectly the appearance which *carburetted hydrogen gas* or *fire damp* presents when burnt in day-light. But the greatest issue of gas was in the rivulet itself, distant about twenty yards from the place where the gas was burning. The rivulet when I visited the place was swollen and muddy, so as to prevent its bottom from being seen. But the gas issued up through it in one place with great violence, as if it had been in a state of compression under the surface of the earth; and the thickness of the jet could not be less than two or three inches in diameter. We set the gas on fire as it issued through the water. It burnt for some time with a good deal of splendour; but as the rivulet was swollen and rushing along with great impetuosity, the regularity of the issue was necessarily disturbed, and the gas was extinguished." Dr Thomson, found this gas to consist of *two* volumes of hydrogen gas and *one* volume of vapour of carbon; and as its specific gravity was 0.555, and as it issues in great abundance, he remarks that it might be used for filling air-balloons. " Were we assured, " he adds, " that it would continue to issue in as great abundance as at present, it might be employed in lighting the streets of Glasgow." *

A very curious natural phenomenon, called the *Lantern* or *Natural Lighthouse* of Maracaybo, has been witnessed in South America. A bright light is seen every night on a mountainous and uninhabited spot on the banks of the river Catatumbo, near its junction with the Sulia. It is easily distinguished at a greater distance than *forty* leagues, and as it is nearly in the meridian of the opening of the Lake of Maracaybo, navigators are guided by it as by a lighthouse. This phenomenon is not only seen from the sea coast but also from the interior of the country,—at Merida, for example, where M. Palacios observed it for two years. Some persons have ascribed this remarkable phenomenon to a thunder storm, or to electrical explosions, which might take place daily in a pass in the mountains; and it has even been asserted, that the rolling of thunder is heard by those who approach the spot. Others suppose it to be an air volcano, like those on the Caspian Sea, and that it is caused by asphaltic soils like those of Mena. It is more probable, however, that it is a sort of carburetted hydrogen, as hydrogen gas is disengaged from the ground in the same district. †

* Edinburgh Journal of Science, No. i. New Series, p. 71. —75.

† Humboldt's Personal Narrative, Vol. iv. p. 254, note.

Grand as the chemical operations are which are going on in the great laboratory of Nature, and alarming as their effects appear when they are displayed in the terrors of the earthquake and the volcano, yet they are not more wonderful to the philosopher than the minute though analogous operations which are often at work near our own persons, unseen and unheeded. It is not merely in the bowels of the earth that highly expansive elements are imprisoned and restrained, and occasionally called into tremendous action by the excitation of heat and other causes. Fluids and vapours of a similar character exist in the very gems and precious stones which science has contributed to luxury and to the arts.

In examining with the microscope the structure of mineral bodies, I discovered in the interior of many of the gems thousands of cavities of various forms and sizes. Some had the shape of hollow and regularly formed crystals : others possessed the most irregular outline, and consisted of many cavities and branches united without order, but all communicating with each other. These cavities sometimes occurred singly, but most frequently in groups forming strata of cavities at one time perfectly flat and at another time curved. Several such strata were often found in the same specimen, sometimes parallel to each other, at other times inclined, and forming all varieties of angles with the faces of the original crystal.

These cavities, which occurred in *sapphire, chrysoberyl, topaz, beryl, quartz, amethyst, peridot,* and other substances, were sometimes sufficiently large to be distinctly seen by the naked eye, but

most frequently they were so small as to require a
high magnifying power to be well seen, and often
they were so exceedingly minute, that the highest
magnifying powers were unable to exhibit their
outline.

The greater number of these cavities, whether large
or small, contain two new fluids different from any
hitherto known, and possessing remarkable physical properties.   These two fluids are in general
perfectly transparent and colourless, and they exist
in the same cavity in actual contact, without mixing together in the slightest degree.   One of them
expands *thirty* times more than water, and at a temperature of about 80° of Fahrenheit it expands so as
to fill up the vacuity in the cavity.   This will be understood from the annexed figure, where A B C D is

Fig. 75.

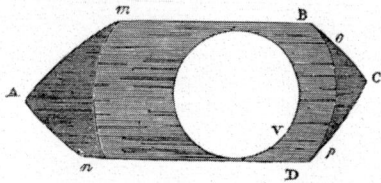

the cavity, *m n p o* the highly expansible fluid in
which at low temperatures there is always a vacuity V,
like an air-bubble in common fluids, and *A m n, C o p,*
the second fluid occupying the angles A and C.
When heat such as that of the hand is applied to
the specimen, the vacuity V gradually contracts in
size, and wholly vanishes at a temperature of about

80°, as shewn in Fig. 76. The fluids are shaded, as
in these two figures, when they are seen by light
reflected from their surfaces.

Fig. 76.

When the cavities are large, as in Fig 77, compared with the quantity of expansible fluid *m n p o*,
the heat converts the fluid into vapour, an effect

Fig. 77.

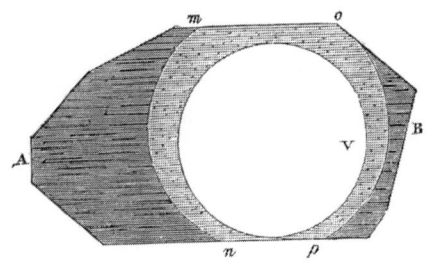

which is shewn by the circular cavity V becoming
larger and larger till it fills the whole space *m n
o p.*
When any of these cavities, whether they are

filled with fluid or with vapour, is allowed to cool, the vacuity V reappears at a certain temperature. In the fluid cavities the fluid contracts, and the small vacuity appears, which grows larger and larger till it resumes its original size. When the cavities are large several small vacuities make their appearance and gradually unite into one, though they sometimes remain separate. In deep cavities a very remarkable phenomenon accompanies the reappearance of the vacuity. At the instant that the fluid has acquired the temperature at which it quits the sides of the cavity, an effervescence or rapid ebullition takes place, and the transparent cavity is for a moment opaque, with an infinite number of minute vacuities, which instantly unite into one that goes on enlarging as the temperature diminishes. In the vapour cavities the vapour is reconverted by the cold into fluid, and the vacuity V, Fig 77, gradually contracts till all the vapour has been precipitated. It is curious to observe, when a great number of cavities are seen at once in the field of the microscope, that the vacuities all disappear and reappear at the same instant.

While all these changes are going on in the expansible fluid, the other denser fluid at A and C, Fig. 75, 76, remains unchanged either in its form or magnitude. On this account I experienced considerable difficulty in proving that it was a fluid. The improbability of two fluids existing in a transparent state in absolute contact, without mixing in the slightest degree, or acting upon each other, induced many persons to whom I shewed the phenomenon to consider the lines m n, o p, Fig. 75, 76, as a partition in the cavity, or the spaces A m n, C o p, either as filled with solid matter, or as corners into which the expanding fluid would not penetrate. The regular curvature, however, of the boundary line *m n*, *o p*, and other facts, rendered these suppositions untenable.

This difficulty was at last entirely removed by the discovery of a cavity of the form shewn in the annexed figure, where A, B, and C, are three portions of the expansible fluid separated by the inter-

Fig. 78.

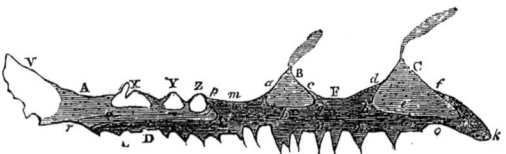

position of the second fluid D E F.   The first portion A of the expansible fluid had four vacuities V, X, Y, Z, while the other two portions B, C, had no vacuity.   In order to determine if the vacuities of the portions B, C, had passed over to A, I took an accurate drawing of the appearances at a temperature of 50°, as shewn in the figure, and I watched the changes which took place in raising the temperature to 83°.   The portion A gradually expanded itself till it filled up all the four vacuities V, X, Y, and Z, but as the vacuities B, C, had no vacuities, they could expand themselves only by pushing back the supposed second fluid D E F. This effect actually took place.   The dense fluid quitted the side of the cavity at F.   The two portions B, C, of the expansible fluid instantly united, and

the dense fluid having retreated to the limit *m n*
n o, its other limit advanced to *p q r*, thus proving
it to be a real fluid.   This experiment, which I
have often shewn to others, involves one of those
rare combinations of circumstances which nature
sometimes presents to us in order to illustrate her
most mysterious operations.   Had the portions
B, C, been accompanied, as is usual, with their vacuities, the interposed fluid would have remained
immoveable between the two equal and opposite
expansions ; but owing to the accidental circumstance of these vacuities having passed over into
the other branch A of the cavity, the fluid yielded to the difference of the expansive forces between
which it lay, and thus exhibited its fluid character
to the eye.

When we examine these cavities narrowly, we
find that they are actually little laboratories, in

Fig. 79.

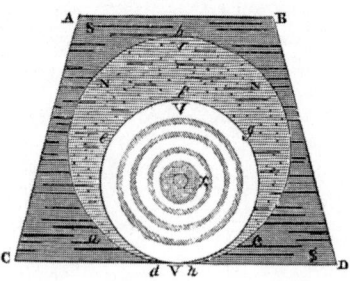

which chemical operations are constantly going on,
and beautiful optical phenomena continually dis-

playing themselves. Let A B D C, for example, be the summit of a crystallized cavity in topaz, S S representing the dense, N N the expansible fluid, bounded by a circular line *a b c d*, and V V the vacuity in the new fluid, bounded by the circle *e f g h*. If the face A B D C is placed under a compound microscope, so that light may be reflected at an angle less than that of total reflexion, and if the observer now looks through the microscope, the temperature of the room being 50°, he will see the second fluid S S shining with a very feeble reflected light, the dense fluid N N with a light perceptibly brighter, and the vacuity V V, with a light of considerable brilliancy. The boundaries *a b c d*, *e f g h* are marked by a well-defined outline, and also by the concentric coloured rings of thin plates produced by the extreme thinness of each of the fluids at their edges.

If the temperature of the room is raised slowly to 58°, a brown spot will appear at *x* in the centre of the vacuity V V. This spot indicates the commencement of evaporation from the expansible fluid below, and arises from the partial precipitation of the vapour in the roof of the cavity. As the heat increases, the brown spot enlarges and becomes very dark. It is then succeeded by a white spot, and one or more coloured rings rise in the centre of the vacuity. The vapour then seems to form a drop, and all the rings disappear by retiring to the centre, but only to reappear with new lustre. During the application of heat, the circle *e f g h* contracts and dilates like the pupil of the eye. When the vaporization is so feeble as to produce only a single ring of one or two tints of the second

Y

order, they vanish instantly by breathing upon the crystal, but when the slight heat of the breath reaches the fluid, it throws off fresh vapour, and the rings again appear.

If a drop of ether is put upon the crystal when the rings are in a state of rapid play, the cold produced by its evaporation causes them to disappear, till the temperature again rises. When the temperature is perfectly uniform, the rings are stationary, as shewn between V and V in Fig. 79; and it is interesting to observe the first ring produced by the vapour swelling out to meet the first ring at the margin of the fluid, and sometimes coming so near it that the darkest parts of both form a broad black band. As the heat increases, the vacuity V V diminishes and disappears at 79°, exhibiting many curious phenomena, which we have not room to describe.

Having fallen upon a method of opening the cavities, and looking at the fluids, I was able to examine their properties with more attention. When the expansible fluid first rises from the cavity upon the surface of the topaz, it neither remains still like the fixed oils, nor disappears like evaporable fluids. Under the influence, no doubt, of heat and moisture it is in a state of constant motion, now spreading itself on a thin plate over a large surface, and now contracting itself into a deeper and much less extended drop. These contractions and extensions are marked by very beautiful optical phenomena. When the fluid has stretched itself out into a thin plate, it ceases to reflect light like the thinest part of the soap-bubble, and when it is again accumulated into a thicker drop, it is covered with the coloured rings of thin plates.

After performing these motions, which sometimes last for ten minutes, the fluid suddenly disappears, and leaves behind it a sort of granular residue. When examining this with a single microscope, it again started into a fluid state, and extended and contracted itself as before. This was owing to the humidity of the hand which held the microscope, and I have been able to restore by moisture the fluidity of these grains twenty days after they were formed from the fluid. This portion was shewn to the Reverend Dr Fleming, who remarked, that, had he observed it accidentally, he would have ascribed its apparent vitality to the movements of some of the animals of the genus Planaria.

After the cavity has remained open for a day or two, the dense fluid comes out and quickly hardens into a transparent and yellowish resinous-looking substance, which absorbs moisture, though with less avidity than the other. It is not volatilized by heat, and is insoluble in water and alcohol. It readily dissolves, however, with effervescence in the sulphuric, nitric, and muriatic acids. The residue of the expansible fluid is volatilized by heat, and is dissolved, but without effervescence, in the above-mentioned acids. The refractive power of the dense fluid is about 1.295, and of the expansible one 1.131.

The particles of the dense fluid have a very powerful attraction for each other and for the mineral which contains them, while those of the expansible fluid have a very slight attraction for one another, and also for the substance of the mineral. Hence the two fluids never mix, the dense fluid being attracted to the angles of angular cavities, or filling the narrow necks by which two cavities communicate. The expansible fluid, on the other hand, fills the wide parts of the cavities, and in deep and round cavities it lies above the dense fluid.

When the dense fluid occupies the necks which join two cavities, it performs the singular function of a fluid valve, opening and shutting itself according to the expansions or contractions of the other fluid. The *fluid valves* thus exhibited in action may suggest some useful hints to the mechanic and the philosopher, while they afford ground of curious speculation in reference to the functions of animal and vegetable bodies. In the larger organizations of ordinary animals, where gravity must in general overpower, or at least modify, the influence of capillary attraction, such a mechanism is neither necessary nor appropriate ; but, in the lesser functions of the same animals, and in almost all the microscopic structures of the lower world, where the force of gravity is entirely subjected to the more powerful energy of capillary forces, it is extremely probable that the mechanism of immiscible fluids and fluid valves is generally adopted.

In several cavities in minerals I have found crystallised and other bodies, sometimes transparent crystals, sometimes black spicular crystals, and sometimes black spheres, all of which are moveable within the cavity. In some cavities the two new fluids occur in an indurated state, and others I have found to be lined with a powdery matter. This last class of cavities occurred in topaz, and they were distinguished from all others by the extraordinary beauty and symmetry of their form. One of these cavities represented a finely ornamented sceptre, and, what is still more singular, the different parts of which it is composed lay in different planes.

When the gem which contains the highly expansive fluid is strong, and the cavity not near the surface, heat may be applied to it without danger, but in the course of my experiments on this subject, the mineral has often burst with a tremendous explosion, and in one case wounded me on the brow. An accident of the same kind occurred to a gentleman, who put a crystal into his mouth for the purpose of expanding the fluid. The specimen burst with great force and cut his mouth, and the fluid which was discharged from the cavity had a very disagreeable taste.

In the gems which are peculiarly appropriated for female ornaments, cavities containing the expansible fluid frequently occur, and if these cavities should happen to be very near the surface or the edge of the stone, the fever heat of the body might be sufficient to burst them with an alarming and even dangerous explosion. I have never heard of any such accident having occurred; but if it has, or if it ever shall occur, and if its naturally marvellous character shall be heightened by any calamitous results, the phenomena described in the preceding pages will strip it of its wonder.

There are no facts in chemistry more interesting than those which relate to the changes of colour, which are produced by the mixture of fluids, and to the creation of brilliant colours by the combination of bodies in which no colouring matter is visible.

Feats of this kind are too common and too generally known to require to be noticed in a work like this. The art of producing such changes was known to some of the early impostors, who endeavoured to obtain a miraculous sanction to their particular dogmas. Marcos, the head of one of the sects that wished to engraft paganism upon Christianity, is said to have filled three transparent glasses with white wine, and while he prayed, the wine in one of the glasses became red like blood, that in another became purple, and that in the third sky-blue. Such transformations present no difficulty to the chemist. There are several fluids, such as some of the coloured juices of plants, which change their colour rapidly and without any additional ingredient; and in other cases, there would be no difficulty in making additions to fluids, which should produce a change of colour at any required instant.

A very remarkable experiment of an analogous nature has been publicly exhibited in modern times. Professor Beyruss, who lived at the court of the Duke of Brunswick, one day pronounced to his highness that the dress which he wore should during dinner become red; and the change actually took place to the astonishment of the prince and the rest of his guests. M. Vogel, who has recorded this curious fact, has not divulged the secret of the German chemist; but he observes, that, if we pour lime-water into the juice of beet root, we shall obtain a colourless liquid; and that a piece of white cloth dipped in this liquid and dried rapidly will in a few hours become red by the mere contact of air. M. Vogel is also of opinion that this singular effect would be accelerated in an apartment where champagne or other fluids charged with carbonic acid are poured out in abundance.

Among the wonders of chemistry we must number the remarkable effects produced upon the human frame by the inhalation of *paradise* or *intoxicating gas*, as it has been called. This gas is known to chemists by the name of the *nitrous oxide*, or the *gaseous oxide of azote*, or the *protoxide of nitrogen*. It differs from atmospheric air only in the proportion of its ingredients, atmospheric air being composed of twenty-seven parts of oxygen and seventy-three of nitrogen, while the nitrous oxide consists of thirty-seven parts of oxygen, and sixty-seven of nitrogen. The most convenient way of procuring the gas is to expose nitrate of ammonia in a tubulated glass retort to the heat of an argand's lamp between 400° and 500° of Fahrenheit. The salt first melts ; bubbles of gas begin to rise from the mass, and in a short time a brisk effervescence takes place, which continues till all the salt has disappeared. The products of this operation are the nitrous oxide and water, the watery vapour being condensed in the neck of the retort while the gas is received over water. The gas thus obtained is generally white, and hence, when it is to be used for the purposes of respiration, it should remain at least an hour over water, which will absorb the small quantity of acid and of nitrate of ammonia which adhere to it. A pound of the nitrate of ammonia will in this way yield five cubic feet of gas fit for the purpose of inhalation.

It was discovered by Sir Humphry Davy, that this gas could be safely taken into the lungs, and that it was capable of supporting respiration for a few minutes. In making this experiment he was surprised to find that it produced a singular species of intoxication, which he thus describes : " I breathed," says he, " three quarts of oxide from and into a silk bag for more than half a minute without previously closing my nose or exhausting my lungs. The first inspiration caused a slight degree of giddiness. This was succeeded by an uncommon sense of fulness in the head, accompanied with loss of distinct sensation and voluntary power, a feeling analogous to that produced in the first stage of intoxication, but unattended by pleasurable sensations." In describing the effects of another experiment, he says, " Having previously closed my nostrils and exhausted my lungs, I breathed four quarts of nitrous oxide from and into a silk bag. The first feelings were similar to those produced in the last experiment, but in less than half a minute, the respiration being continued, they diminished gradually, and were succeeded by a highly pleasurable thrilling, particularly in the chest and the extremities. The objects around me became dazzling, and my hearing more acute. Towards the last respiration the thrilling increased, the sense of muscular power became greater, and at last an irresistible propensity to action was indulged in. I recollect but indistinctly what followed ; I knew that my motions were varied and violent. These effects very rarely ceased after respiration. In ten minutes I had recovered my natural state of mind. The thrilling in the extremities continued longer than the other sensations. This experiment was made in the morning ; no languor or exhaustion was consequent, my feelings through the day were

as usual, and I passed the night in undisturbed repose."

In giving an account of another experiment with
this gas, Sir Humphry thus describes his feelings :
" Immediately after my return from a long journey,
being fatigued, I respired nine quarts of nitrous
oxide, having been precisely thirty-three days without breathing any. The feelings were different
from those I had experienced on former experiments. After the first six or seven respirations,
I gradually began to lose the perception of external
things, and a vivid and intense recollection of some
former experiments passed through my mind, so
that I called out, ' what an annoying concatenation
of ideas.' "

Another experiment made by the same distinguished chemist was attended by still more remarkable
results. He was shut up in an air-tight breathing
box, having a capacity of about nine and a-half cubic
feet, and he allowed himself to be habituated to the
excitement of the gas, which was gradually introduced. After having undergone this operation for
an hour and a quarter, during which eighty quarts
of gas were thrown in, he came out of the box and
began to respire twenty quarts of unmingled nitrous
oxide. " A thrilling," says he, " extending from the
chest to the extremities, was almost immediately
produced. I felt a sense of tangible extension,
highly pleasurable in every kind, my visible impressions were dazzling, and apparently magnified.
I heard distinctly every sound in the room, and
was perfectly aware of my situation. By degrees,
as the pleasurable sensation increased, I lost all
connection with external things ; trains of vivid

visible images rapidy passed through my mind, and
were connected with words in such a manner as to
produce perceptions perfectly novel.   I existed in
a world of newly connected and newly modified
ideas.   When I was awakened from this same de-
lirious trance by Dr Kinglake, who took the bag
from my mouth, indignation and pride were the
first feelings produced by the sight of the persons
about me.   My emotions were enthusiastic and
sublime, and for a moment I walked round the
room, perfectly regardless of what was said to me.
As I recovered my former state of mind, I felt an
inclination to communicate the discoveries I had
made during the experiment.   I endeavoured to
recal the ideas ; they were feeble and indistinct.
One recollection of terms, however, presented itself,
and with the most intense belief and prophetic
manner I exclaimed to Dr Kinglake, ' nothing exists but thoughts ; the universe is composed of impressions, ideas, pleasures, and pains.' "

These remarkable properties induced several persons to repeat the experiment of breathing this ex-
hilarating medicine.   Its effects were, as might have
been expected, various in different individuals ; but
its general effect was to produce in the gravest and
most phlegmatic the highest degree of exhilaration
and happiness, unaccompanied with languor or depression.   In some it created an irresistible disposition to laugh, and in others a propensity to muscular
exertion.   In some it impaired the intellectual
functions, and in several it had no sensible effect,
even when it was breathed in the purest state and in
considerable quantities.   It would be an inquiry of
no slight interest to ascertain the influence of this

gas over persons of various bodily temperaments, and upon minds varying in their intellectual and moral character.

Although Sir Humphry Davy experienced no unpleasant effects from the inhalation of the nitrous oxide, yet such effects are undoubtedly produced; and there is reason to believe that even permanent changes in the constitution may be induced by the operation of this remarkable stimulant. Two very interesting cases of this kind presented themselves to Professor Silliman of Yale College, when the nitrous oxide was administered to some of his pupils. The students had been in the habit for several years of preparing this gas and administering it to one another, and these two cases were the only remarkable ones which deserved to be recorded. We shall describe them in Professor Silliman's own words :—

" A gentleman about nineteen years of age, of a sanguine temperament, and cheerful temper, and in the most perfect health, inhaled the usual quantity of the nitrous oxide when prepared in the ordinary manner. Immediately his feelings were uncommonly elevated, so that, as he expressed it, he could not refrain from dancing and shouting. Indeed to such a degree was he excited, that he was thrown into a frightful fit of delirium, and his exertions became so violent, that after a while he sunk to the earth exhausted, and there remained, until having by quiet in some degree recovered his strength, he again arose only to renew the most convulsive muscular efforts, and the most piercing screams and cries ; within a few moments, overpowered by the intensity of the paroxysm, he again fell to the ground apparently senseless and panting vehemently. The long continuance and violence of the affection alarmed his companions, and they ran for professional assistance. They were, however, encouraged by the person to whom they applied to hope that he would come out of his trance without injury, but for the space of two hours these symptoms continued; he was perfectly unconscious of what he was doing, and was in every respect like a maniac. He states, however, that his *feelings vibrated* between perfect happiness and the most consummate misery. In the course of the afternoon, and after the first violent effects had subsided, he was compelled to lie down two or three times, from excessive fatigue, although he was immediately aroused upon any one's entering the room. The effects remained in a degree for three or four days accompanied by a hoarseness, which he attributed to the exertion made while under the immediate influence of the gas. This case should produce a degree of caution, especially in persons of a sanguine temperament, whom, much more frequently than others, we have seen painfully and even alarmingly affected."

The other case described by Professor Silliman was that of a man of mature age, and of a grave and respectable character. " For nearly two years previous to his taking the gas his health had been very delicate, and his mind frequently gloomy and depressed. This was peculiarly the case for a few days immediately preceding that time, and his general state of health was such that he was obliged almost entirely to discontinue his studies, and was about to have recourse to medical assistance. In this state of bodily and mental debility he inspired about three quarts of nitrous oxide. The consequences were, an astonishing invigoration of his whole system, and the most exquisite perceptions of delight. These were manifested by an uncommon disposition for pleasantry and mirth, and by extraordinary muscular power. The effects of the gas were felt without diminution for at least thirty hours, and in a greater or less degree for more than a week.

" But the most remarkable effect was that *upon the organs of taste.* Antecedently to taking the gas, he exhibited no peculiar choice in the articles of food, but immediately subsequent to that event, he *manifested a taste for such things only as were sweet,* and for several days *ate nothing but sweet cake.* Indeed this singular taste was carried to such excess, that he used *sugar and molasses not only upon his bread and butter and lighter food, but upon his meat and vegetables.* This he continues to do even at the present time, and although eight weeks have elapsed since he inspired the gas, he is still found *pouring molasses over beef, fish, poultry, potatoes, cabbage, or whatever animal or vegetable food is placed before him.*

" His health and spirits since that time have been uniformly good, and he attributes the restoration of his strength and mental energy to the influence of the nitrous oxide. He is entirely regular in his mind, and now experiences no uncommon exhilaration, but is habitually cheerful, while before he was as habitually grave, and even to a degree gloomy."

Such is a brief and a general account of the principal phenomena of Nature, and the most remarkable deductions of science, to which the name of Natural Magic has been applied. If those who have not hitherto sought for instruction and amusement in the study of the material world shall have found a portion of either in the preceding pages, they will not fail to extend their inquiries to other popular departments of science, even if they are less marked with the attributes of the marvellous. In every region of space, from the infinitely distant recesses of the heavens to the " dark unfathomed caves of ocean," the Almighty has erected monuments of miraculous grandeur, which proclaim the power, the wisdom, and the beneficence of their author. The inscriptions which they bear—the handwriting which shines upon their walls—appeal to the understanding and to the affections, and demand the admiration and the gratitude of every rational being. To remain willingly ignorant of these revelations of the Divine Power is a crime next to that of rejecting the revelation of the Divine will. Knowledge, indeed, is at once the handmaid and the companion of true religion. They mutually adorn and support each other; and beyond the immediate circle of our secular duties, they are the only objects of rational ambition. While the calm deductions of reason regulate the ardour of Christian zeal, the warmth of a holy enthusiasm gives a fixed brightness to the glimmering lights of knowledge.

It is one of the darkest spots in the history of man that these noble gifts have been so seldom combined. In the young mind alone can the two kindred seeds be effectually sown ; and among the improvements which some of our public institutions require, we yet hope to witness a national system of instruction, in which the volumes of Nature and of Revelation shall be simultaneously perused.

D. BREWSTER.

ALLERLY, *April 24th* 1832.

THE END.